Intelligent Systems Reference Library

Volume 168

Series Editors

Janusz Kacprzyk, Polish Academy of Sciences, Warsaw, Poland

Lakhmi C. Jain, Faculty of Engineering and Information Technology, Centre for
Artificial Intelligence, University of Technology, Sydney, NSW, Australia;
Faculty of Science, Technology and Mathematics, University of Canberra,
Canberra, ACT, Australia;
KES International, Shoreham-by-Sea, UK;
Liverpool Hope University, Liverpool, UK

The aim of this series is to publish a Reference Library, including novel advances and developments in all aspects of Intelligent Systems in an easily accessible and well structured form. The series includes reference works, handbooks, compendia, textbooks, well-structured monographs, dictionaries, and encyclopedias. It contains well integrated knowledge and current information in the field of Intelligent Systems. The series covers the theory, applications, and design methods of Intelligent Systems. Virtually all disciplines such as engineering, computer science, avionics, business, e-commerce, environment, healthcare, physics and life science are included. The list of topics spans all the areas of modern intelligent systems such as: Ambient intelligence, Computational intelligence, Social intelligence, Computational neuroscience, Artificial life, Virtual society, Cognitive systems, DNA and immunity-based systems, e-Learning and teaching, Human-centred computing and Machine ethics, Intelligent control, Intelligent data analysis, Knowledge-based paradigms, Knowledge management, Intelligent agents, Intelligent decision making, Intelligent network security, Interactive entertainment, Learning paradigms, Recommender systems, Robotics and Mechatronics including human-machine teaming, Self-organizing and adaptive systems, Soft computing including Neural systems, Fuzzy systems, Evolutionary computing and the Fusion of these paradigms, Perception and Vision, Web intelligence and Multimedia.

** Indexing: The books of this series are submitted to ISI Web of Science, SCOPUS, DBLP and Springerlink.

More information about this series at http://www.springer.com/series/8578

Seiki Akama · Yasuo Kudo ·
Tetsuya Murai

Topics in Rough Set Theory

Current Applications to Granular Computing

 Springer

Seiki Akama
C-Republic
Kawasaki, Japan

Yasuo Kudo
Muroran Institute of Technology
Muroran, Japan

Tetsuya Murai
Chitose Institute of Science and Technology
Chitose, Hokkaido, Japan

ISSN 1868-4394 ISSN 1868-4408 (electronic)
Intelligent Systems Reference Library
ISBN 978-3-030-29568-4 ISBN 978-3-030-29566-0 (eBook)
https://doi.org/10.1007/978-3-030-29566-0

This Springer imprint is published by the registered company Springer Nature Switzerland AG
The registered company address is: Gewerbestrasse 11, 6330 Cham, Switzerland

Foreword

This book presents the fundamentals of rough set theory and the current state of research under that theory, with particular stress on its applications to granular computing. The rough set notion, introduced in the early 1980s by Zdzislaw Pawlak, is a simple but powerful concept used for handling imprecise or incomplete information. The rough set paradigm differs from the well-known fuzzy set concept introduced earlier by Lofti Zadeh in the lack of any concrete numerical measure for membership of an object in such a set. Instead, certain groups of objects are unified and treated as identical from the viewpoint of our knowledge—with such unification representing the (in)accuracy of our data about those objects.

In the original Pawlak's approach, the unification of objects was underpinned by an equivalence relation, the intuition of which stemmed from indistinguishability of objects having the same values of attributes in a given information system. The equivalence classes of such a relation were used to form the lower and upper approximation of a set—the constructs of fundamental importance in rough set theory. Later, the equivalence relation was replaced by more general notions, and eventually by any covering of the universe of objects. Since their advent, rough sets have given rise to a very rich body of research, developed by scholars all over the world and enjoying particular popularity in China. The results of that research have found many practical applications—e.g., in data classification, medical diagnosis, decision theory, control of production processes, etc. An important, fairly recent development is application of rough sets to granular computing—which is the specially important topic of this book.

The book gives the reader a rich, orderly arranged overview of rough set theory and its applications. It can play a role of an extensive textbook on rough sets for students, and a valuable tool for scholars and practitioners who have to deal with imperfect, imprecise, or absent information. And as nowadays such information is ubiquitous in our professional and even private lives, in my opinion everybody can profit in diverse ways from reading this book.

Warsaw, Poland
July 2019

Beata Konikowska

Preface

In 1982, Pawlak proposed rough set theory to formalize imprecise and uncertain data and reasoning from data. In rough set theory, objects are grouped together by their features as a homogeneous family, and certain and uncertain members of a set are identified group-wisely by checking whether all members of a group belong to the set or not.

Thus, rough set theory is capable of grouping objects, and it is useful in the analysis of datasets concerning classification. Rough set theory is attracted to those working in various areas, in particular, Artificial Intelligence and computer science.

Later, many workers investigated foundations and applications of rough set theory. As the first stage, it was mainly applied to learning, data analysis, data mining, etc. And rough set theory recently found new applications areas, in particular, *granular computing*. It is an approach to information processing that deals with complex information entities called *granular*. The idea was proposed by Zadeh in 1997, and it is closely related to rough set theory. This book presents current topics of rough set theory related granular computing.

The structure of the book is as follows.

Chapter 1 introduces the backgrounds of rough set theory. We discuss the importance of rough set theory and connections with other related theories. We also give the history of rough set theory. We then provide some comments on later chapters.

Chapter 2 overviews rough set theory. First, we give preliminaries of rough set theory. Second, we give an exposition of algebras for rough set theory. Third, connections of modal logic and rough sets are described. Fourth, rough set logics are introduced. We also consider logics for reasoning about knowledge and logics for knowledge representation. We also give a concise survey of fuzzy logic. Finally, we shortly suggest applications of rough set theory.

Chapter 3 deals with object reduction in rough set theory. We introduce a concept of object reduction that reduces the number of objects as long as possible with keeping the results of attribute reduction in the original decision table.

Chapter 4 introduces a recommendation method by directly setting of users' preference patterns of items for similarity evaluation between users. Our method for extraction of preference patterns is based on partial pairwise comparison of items and it is applicable to represent "cyclic" preference patterns. A tourist spot recommender system is implemented as a prototype of recommender system based on the proposed approach and the implemented recommender system is evaluated by experiments.

Chapter 5 discusses rough-set-based interrelationship mining for incomplete decision tables. Rough-set-based interrelationship mining enables to extract characteristics by comparing the values of the same object between different attributes. To apply this interrelationship mining to incomplete decision tables with null values, in this study, we discuss the treatment of null values in interrelationships between attributes. We introduce three types of null values for interrelated condition attributes and formulate a similarity relation by such attributes with these null values.

Chapter 6 proposes a parallel computation framework for a heuristic attribute reduction method. Attribute reduction is a key technique to use rough set theory as a tool in data mining. The authors have previously proposed a heuristic attribute reduction method to compute as many relative reducts as possible from a given dataset with numerous attributes. We parallelize our method by using open multiprocessing. We also evaluate the performance of a parallelized attribute reduction method by experiments.

Chapter 7 discusses the heuristic algorithm to computes a relative reduct candidate based on evaluating classification ability of condition attributes. Considering the discernibility and equivalence of objects for condition attributes in relative reducts, we introduce evaluation criteria for condition attributes and relative reducts. The computational complexity of the proposed algorithm is $O(|U|^2|C|^2)$. Experimental results indicate that our algorithm often generates a relative reduct producing a good evaluation result.

Chapter 8, from the viewpoint of approximation, introduces an evaluation criterion for relative reducts using roughness of partitions constructed from them. The outline of relative reduct evaluation we propose is: "Good" relative reducts = relative reducts that provide partitions with approximations as rough and correct as possible. In this sense, we think that evaluation of relative reducts is strictly concerned with evaluation of roughness of approximation.

Chapter 9 concerns recommender systems (RSs), providing personalized information by learning user preferences. User-based collaborative filtering (UBCF) is a significant technique widely utilized in RSs. The traditional UBCF approach selects k-nearest neighbors from candidate neighbors comprised of all users; however, this approach cannot achieve good accuracy and coverage values simultaneously. We present a new approach using covering-based rough set theory to improve traditional UBCF in RSs. In this approach, we insert a user reduction procedure into the traditional UBCF approach. Covering reduction in covering-based rough sets is used to remove redundant users from all users. Then, k-nearest neighbors are selected from candidate neighbors comprised of the reduct-users. Our experimental results suggest that, for the sparse datasets that often occur in real RSs, the proposed approach outperforms the traditional UBCF and can provide satisfactory accuracy and coverage simultaneously.

Chapter 10 applies granular computing to Aristotle's categorical syllogism. Such kind of reasoning is called granular reasoning. For the purpose, two operations called zooming in and out are introduced to reconstruct granules of possible worlds.

Chapter 11 proposes two key concepts—focus and visibility—as modalities of modal logic. Scott-Montague models that we have proposed represent properties of visibility and focus and the concept that p is visible as modal sentence Vp and p is clearly visible-or is in focus-as modal sentence Cp.

Chapter 12 concludes this book by sketching some directions for future work. In the context of soft computing dealing with vagueness, important problems in rough set theory include decision tables, decision rules, attribute reduction, etc. While some of these issues have been discussed in this book, we consider them further. There are several generalizations of rough set theory and unifications with other theories. Thus, we discuss several issues about generalizations and unifications. A theory of three-way decisions is also presented as a promising framework for generalizing rough set theory.

We are grateful to Prof. Lakhmi C. Jain and Prof. Beata Konikowska for useful comments.

Kawasaki, Japan
Muroran, Japan
Chitose, Japan
July 2019

Seiki Akama
Yasuo Kudo
Tetsuya Murai

Contents

Chapter 1
Introduction

Abstract This chapter introduces the backgrounds of rough set theory. We discuss the importance of rough set theory and connections with other related theories. We also give the history of rough set theory. We then provide some comments on later chapters.

Keywords Rough sets · Rough set theory · Imprecise and uncertain data · Reasoning from data

1.1 Backgrounds

A theory to deal with uncertainty, called *rough set theory* was introduced by Pawlak in 1982; see Pawlak [11, 12]. It can be seen as an extension of (standard) *set theory*, in which a subset of a universe is formalized by a pair of sets, i.e., the lower and upper approximations. These approximations can be described by two operators on subsets of the universe.

Observe that, in rough set theory, an *equivalence relation*, i.e., reflexive, symmetric and transitive relation, plays an important role. Based on an equivalence relation, we can define the lower approximation of a given set is the union of all equivalence classes which are subset of the set, and the upper approximation is the union of all equivalence classes which have a non-empty intersection with the set. These approximations can naturally represent incomplete information.

Of course, rough set theory can be developed by relations other than an equivalence relation. But the use of an equivalence relation enables an elegant formalization, and we can obtain simple applications. However, after Pawlak's work, versions of rough set theory using various relations have been proposed in the literature.

Rough set theory is, in particular, helpful in extracting knowledge from data tables and it has been successfully applied to the areas such as data analysis, decision making, machine learning, etc.

We also observe that set theory and logic are strongly connected. This means that rough set-based approaches to knowledge representation and logic-based ones have intimate connections. In fact, *rough set* has several connections with *non-classical*

© Springer Nature Switzerland AG 2020

S. Akama et al., *Topics in Rough Set Theory*, Intelligent Systems Reference Library 168, https://doi.org/10.1007/978-3-030-29566-0_1

logics, in particular, modal logic. A lot of work has been done to provide a logical foundation for rough set theory.

In the 1980s, a logic for reasoning about concepts, which is essentially the modal logic S5, was developed based on rough sets by Orlowska [10]. A generalization of rough sets by modal logic using Kripke semantics was also worked out by Yao and Lin [13].

In rough set theory, objects are grouped together by their features as a homogeneous family, and certain and uncertain members of a set are identified group-wisely by checking whether all members of a group belong to the set or not. Thus, rough set theory is capable of grouping objects, and it is useful in the analysis of datasets concerning classification.

We observe that rough set theory may be combined with any other approach to uncertainty. An advantage of the method is that certain and possible rules are processed separately. We can point out that the advantage of rough set theory is that it does not need any preliminary or additional information about data (like prior probability in probability theory, basic probability number in Dempster-Shafer theory, and grade of membership in fuzzy set theory).

Now, rough set theory becomes one of the most important frameworks for imprecise and uncertain data and reasoning from data. It is also connected with granular computing. In fact, there are many issues on various types of reasoning related to rough set theory.

Later, many workers investigated foundations and applications of rough set theory. As the first stage, it was mainly applied to learning, data analysis, data mining, etc. And rough set theory recently found new applications areas, in particular, *granular computing*. It is an approach to information processing that deals with complex information entities called *granular*. The idea was proposed by Zadeh [14] in 1997, and it is closely related to rough set theory.

Zadeh [14] pointed out that there are basic concepts for human cognition, i.e., granulation, organization and causation. Here, *granulation* involves decomposition of whole into parts, *organization* involves integration of parts into whole, and *causation* involves association of causes with effects. Granulation of an object A leads to a collection of granules of A, with a granule being a clump of points (objects) drawn together by indistinguishability, similarity, proximity or functionality.

Thus, we can give the definition of granular computing as follows: granular computing is a formal framework for computing which handles granular and its reasoning. Although Zadeh sketched a theory of information granulation by means of fuzzy logic, there are other approaches to granular computing as Zadeh suggested. If we seriously consider indistinguishability in the context of granular, we believe that rough set theory can obviously serve as a foundation for granular computing. This book presents current topics of rough set theory, including its applications to granular computing.

1.2 About This Book

We have already published a book on reasoning about rough sets [1] in 2018, but the present book focusses on specific topics which are worked out by the authors and interested in researchers in the area. The rest of this book is as follows:

Chapter 2 overviews rough set theory. First, we give preliminaries of rough set theory. Second, we give an exposition of algebras and logics for rough set theory. Third, connections of modal logic and rough sets are described. Fourth, rough set logics are introduced. We also consider logics for reasoning about knowledge and logics for knowledge representation. We also give a concise survey of fuzzy logic. Finally, we shortly suggest applications of rough set theory.

Chapter 3 deals with object reduction in rough set theory. We introduce a concept of object reduction that reduces the number of objects as long as possible with keeping the results of attribute reduction in the original decision table.

Chapter 4 introduces a recommendation method by directly setting of users' preference patterns of items for similarity evaluation between users. Our method for extraction of preference patterns is based on partial pairwise comparison of items and it is applicable to represent "cyclic" preference patterns. A tourist spot recommender system is implemented as a prototype of recommender system based on the proposed approach and the implemented recommender system is evaluated by experiments.

Chapter 5 discusses rough-set-based interrelationship mining for incomplete decision tables. Rough-set-based interrelationship mining enables to extract characteristics by comparing the values of the same object between different attributes. To apply this interrelationship mining to incomplete decision tables with null values, in this study, we discuss the treatment of null values in interrelationships between attributes. We introduce three types of null values for interrelated condition attributes and formulate a similarity relation by such attributes with these null values.

Chapter 6 proposes a parallel computation framework for a heuristic attribute reduction method. Attribute reduction is a key technique to use rough set theory as a tool in data mining. The authors have previously proposed a heuristic attribute reduction method to compute as many relative reducts as possible from a given dataset with numerous attributes. We parallelize our method by using open multiprocessing. We also evaluate the performance of a parallelized attribute reduction method by experiments.

Chapter 7 discusses the heuristic algorithm to computes a relative reduct candidate based on evaluating classification ability of condition attributes. Considering the discernibility and equivalence of objects for condition attributes in relative reducts, we introduce evaluation criteria for condition attributes and relative reducts. The computational complexity of the proposed algorithm is $O(|U|^2|C|^2)$. Experimental results indicate that our algorithm often generates a relative reduct producing a good evaluation result.

Chapter 8, from the viewpoint of approximation, introduces an evaluation criterion for relative reducts using roughness of partitions constructed from them. The outline of relative reduct evaluation we propose is: "Good" relative reducts = relative reducts that provide partitions with approximations as rough and correct as possible. In this sense, we think that evaluation of relative reducts is strictly concerned with evaluation of roughness of approximation.

Chapter 9 concerns Recommender systems (RSs), providing personalized information by learning user preferences. User-based collaborative filtering (UBCF) is a significant technique widely utilized in RSs. The traditional UBCF approach selects k-nearest neighbors from candidate neighbors comprised by all users; however, this approach cannot achieve good accuracy and coverage values simultaneously. We present a new approach using covering-based rough set theory to improve traditional UBCF in RSs. In this approach, we insert a user reduction procedure into the traditional UBCF approach. Covering reduction in covering-based rough sets is used to remove redundant users from all users. Then, k-nearest neighbors are selected from candidate neighbors comprised by the reduct-users. Our experimental results suggest that, for the sparse datasets that often occur in real RSs, the proposed approach outperforms the traditional UBCF, and can provide satisfactory accuracy and coverage simultaneously.

Chapter 10 applies granular computing to Aristotle's categorical syllogism. Such kind of reasoning is called granular reasoning. For the purpose, two operations called zooming in and out are introduced to reconstruct granules of possible worlds.

Chapter 11 proposes two key concepts-focus and visibility-as modalities of modal logic. Scott-Montague models that we have proposed represent properties of visibility and focus and the concept that p is visible as modal sentence Vp and p is clearly visible-or is in focus-as modal sentence Cp.

Chapter 12 concludes this book by sketching some directions for future work. In the context of soft computing dealing with vagueness, important problems in rough set theory include decision tables, decision rules, attribute reduction, etc. While some of these issues have been discussed in this book, we consider them further. There are several generalizations of rough set theory and unifications with other theories. Thus, we discuss several issues about generalizations and unifications. A theory of three-way decisions is also presented as a promising framework for generalizing rough set theory.

Chapter 2 is based on some chapters included in Akama et al. [1]. Chapter 3 is a revised version of Kudo and Murai [7], and Chap. 4 is based on Kudo et al. [8]. Chapter 5 is a revised version of Kudo and Murai [6]. Chapter 6 revises Kudo and Murai [5]. Chapter 7 is based on Kudo and Murai [4]. Chapter 8 uses the material in Kudo and Murai [3] and Chap. 9 is based on Zhang et al. [15]. Chapter 10 is a revised version of Murai et al. [9]. Chapter 11 uses the material in Kudo and Murai [2]. Chapter 12 is a new paper written for the present book.

References

1. Akama, S., Murai, T., Kudo, Y.: Reasoning with Rough Sets. Springer, Heidelberg (2018)
2. Kudo, Y., Murai, T.: A modal characterization of visibility and focus in granular reasoning. JACIII **13**, 297–303 (2009)
3. Kudo, Y., Murai, T.: An evaluation method of relative reducts based on roughness of partitions. IJCiNi **4**, 50–62 (2010)
4. Kudo, Y., Murai, T.: Heuristic algorithm for attribute reduction based on classification ability by condition attributes. J. Adv. Comput. Intell. Intell. Inform. **15**, 102–109 (2011)
5. Kudo, Y., Murai, T.: A parallel computation method for heuristic attribute reduction using reduced decision tables. JACIII **17**, 371–376 (2013)
6. Kudo, Y., Murai, T.: Rough-set-based interrelationship mining for incomplete decision tables. JACIII **20**, 712–720 (2016)
7. Kudo, Y., Murai, T.: An attempt of object reduction in rough set theory. Proc. SCIS & ISIS **2018**, 33–36 (2018)
8. Kudo, Y., Kuroda, M., Murai, T.: Proposal of a recommendation method by direct setting of preference patterns based on interrelationship mining. Proc. ISASE-MAICS (2018)
9. Murai, T., Sato, Y., Resconi, G., Nakata, M.: Granular reasoning using zooming in & out: Part 2. Aristotle's categorial syllogism. Electron. Notes Theor. Comput. Sci. **82**, 186–197 (2003)
10. Orlowska, E.: Kripke models with relative accessibility relations and their applications to inferences from incomplete information. In: Mirkowska, G., Rasiowa, H. (eds.) Mathematical Problems in Computation Theory, pp. 327–337. Polish Scientific Publishers, Warsaw (1987)
11. Pawlak, P.: Rough sets. Int. J. Comput. Inf. Sci. **11**, 341–356 (1982)
12. Pawlak, P.: Rough Sets: Theoretical Aspects of Reasoning about Data. Kluwer, Dordrecht (1991)
13. Yao, Y., Lin, T.: Generalization of rough sets using modal logics. Intell. Autom. Soft Comput. **2**, 103–120 (1996)
14. Zadeh, L.: Toward a theory of fuzzy information granulation and its centrality in human reasoning and fuzzy logic. Fuzzy Sets Syst. **90**, 11–127 (1997)
15. Zhang, Z., Kudo, Y., Murai, T.: Neighbor selection for user-based collaborative filtering using covering-based rough sets. Ann. Oper. Res. **256**, 359–374 (2016)

Chapter 2
Overview of Rough Set Theory

Abstract This chapter overviews rough set theory. First, we give preliminaries of rough set theory. Second, we give an exposition of algebras for rough set theory. Third, connections of modal logic and rough sets are described. Fourth, rough set logics are introduced. We also consider logics for reasoning about knowledge and logics for knowledge representation. We also give a concise survey of fuzzy logic. Finally, we shortly suggest applications of rough set theory.

Keywords Rough set theory · Algebras · Logics for rough sets

2.1 Rough Sets

In 1982, Pawlak introduced a concept of *rough set* to provide a theory of knowledge and classification (Pawlak [47], Akama et al. [4]). Here, we concisely review it. By *object*, we mean anything we can think of, for example, real things, states, abstract concepts, etc. We can assume that knowledge is based on the ability to classify objects. Thus, knowledge is necessarily connected with the variety of classification patterns related to specific parts of the real or abstract world, called the *universe of discourse* (or the universe).

We assume the usual notation for set theory. Let U be non-empty finite set of objects called the *universe of discourse*. Any subset $X \subseteq U$ of the universe is called a *concept* or a *category* in U. Any family of concepts in U is called *knowledge* about U. Note that the empty set $\emptyset$ is also a concept.

We mainly deal with concepts which form a partition (classification) of a certain universe U, i.e. in families $C = \{X_1, X_2, \ldots, X_n\}$ such that $X_i \subseteq U, X_i \neq \emptyset, X_i \cap X_j = \emptyset$ for $i \neq j, i, j = 1, \ldots, n$ and $\bigcup X_i = U$. A family of classifications over U is called a *knowledge base* over U.

Classifications can be specified by using *equivalence relations*. If R is an equivalence relation over U, then U/R means the family of all equivalence classes of R (or classification of U) referred to as categories or concepts of R. $[x]_R$ denotes a category in R containing an element $x \in U$.

© Springer Nature Switzerland AG 2020
S. Akama et al., *Topics in Rough Set Theory*, Intelligent Systems
Reference Library 168, https://doi.org/10.1007/978-3-030-29566-0_2

A *knowledge base* is defined as a relational system, $K = (U, \mathbf{R})$, where $U \neq \emptyset$ is a finite set called the *universe*, and $\mathbf{R}$ is a family of equivalence relations over U. $IND(K)$ means the family of all equivalence relations defined in K, i.e., $IND(K) = \{IND(\mathbf{P}) \mid \emptyset \neq \mathbf{P} \subseteq \mathbf{R}\}$. Thus, $IND(K)$ is the minimal set of equivalence relations, containing all elementary relations of K, and closed under set-theoretical intersection of equivalence relations.

If $\mathbf{P} \subseteq \mathbf{R}$ and $\mathbf{P} \neq \emptyset$, then $\bigcap \mathbf{P}$ denotes the intersection of all equivalence relations belonging to $\mathbf{P}$, denoted $IND(\mathbf{P})$, called an *indiscernibility relation* of $\mathbf{P}$. It is also an equivalence relation, and satisfies:

$$[x]_{IND(\mathbf{P})} = \bigcap_{R \in \mathbf{P}} [x]_R.$$

Thus, the family of all equivalence classes of the equivalence relation $IND(\mathbf{P})$, i.e., $U/IND(\mathbf{P})$ denotes knowledge associated with the family of equivalence relations $\mathbf{P}$. For simplicity, we will write $U/\mathbf{P}$ instead of $U/IND(\mathbf{P})$.

$\mathbf{P}$ is also called $\mathbf{P}$-*basic knowledge*. Equivalence classes of $IND(\mathbf{P})$ are called *basic categories (concepts)* of knowledge $\mathbf{P}$. In particular, if $Q \in \mathbf{R}$, then Q is called a Q-*elementary knowledge* (about U in K) and equivalence classes of Q are referred to as Q-*elementary concepts (categories)* of knowledge $\mathbf{R}$.

Now, we describe the fundamentals of rough sets. Let $X \subseteq U$ and R be an equivalence relation. We say that X is R-*definable* if X is the union of some R-basic categories; otherwise X is R-*undefinable*.

The R-definable sets are those subsets of the universe which can be exactly defined in the knowledge base K, whereas the R-undefinable sets cannot be defined in K. The R-definable sets are called R-*exact sets*, and R-undefinable sets are called R-*inexact* or R-*rough*.

Set $X \subseteq U$ is called *exact* in K if there exists an equivalence relation $R \in IND(K)$ such that X is R-exact, and X is said to be *rough* in K if X is R-rough for any $R \in IND(K)$.

Observe that rough sets can be also defined *approximately* by using two exact sets, referred as a lower and an upper approximation of the set.

Suppose we are given knowledge base $K = (U, \mathbf{R})$. With each subset $X \subseteq U$ and an equivalence relation $R \in IND(K)$, we associate two subsets:

$$\underline{R}X = \bigcup \{Y \in U/R : Y \subseteq X\}$$
$$\overline{R}X = \bigcup \{Y \in U/R : Y \cap X \neq \emptyset\}$$

called the R-*lower approximation* and the R-*upper approximation* of X, respectively. They will be simply called the lower-approximation and the upper-approximation if the context is clear. A pair $(\underline{R}(X), \overline{R}(X))$ is called the *rough set* with respect to X.

It is also possible to define the lower and upper approximation in the following two equivalent forms:

$$\underline{R}X = \{x \in U \mid [x]_R \subseteq X\}$$
$$\overline{R}X = \{x \in U \mid [x]_R \cap X \neq \emptyset\}$$

or

$$x \in \underline{R}X \text{ iff } [x]_R \subseteq X$$
$$x \in \overline{R}X \text{ iff } [x]_R \cap X \neq \emptyset.$$

The above three are interpreted as follows. The set $\underline{R}X$ is the set of all elements of U which can be *certainly* classified as elements of X in the knowledge base R. The set $\overline{R}X$ is the set of elements of U which can be *possibly* classified as elements of X in R.

We define *R-positive region* ($POS_R(X)$), *R-negative region* ($NEG_R(X)$), and *R-borderline region* ($BN_R(X)$) of X as follows:

$$POS_R(X) = \underline{R}X$$
$$NEG_R(X) = U - \overline{R}X$$
$$BN_R(X) = \overline{R}X - \underline{R}X.$$

The positive region $POS_R(X)$ (or the lower approximation) of X is the collection of those objects which can be classified with full certainty as members of the set X, using knowledge R.

The negative region $NEG_R(X)$ is the collection of objects with which it can be determined without any ambiguity, employing knowledge R, that they do not belong to the set X, that is, they belong to the complement of X.

The borderline region $BN_R(X)$ is the set of elements which cannot be classified either to X or to $-X$ in R. It is the undecidable area of the universe, i.e. none of the objects belonging to the boundary can be classified with certainty into X or $-X$ as far as R is concerned.

We show some important results.

Proposition 2.1 *The following hold:*

(1) X is R-definable iff $\underline{R}X = \overline{R}X$
(2) X is rough with respect to R iff $\underline{R}X \neq \overline{R}X$.

Proposition 2.2 shows the basic properties of approximations:

Proposition 2.2 *The R-lower and R-upper approximations satisfy the following properties:*

(1) $\underline{R}X \subseteq X \subseteq \overline{R}X$
(2) $\underline{R}\emptyset = \overline{R}\emptyset = \emptyset, \ \underline{R}U = \overline{R}U = U$
(3) $\overline{R}(X \cup Y) = \overline{R}X \cup \overline{R}Y$
(4) $\underline{R}(X \cap Y) = \underline{R}X \cap \underline{R}Y$
(5) $X \subseteq Y$ implies $\underline{R}X \subseteq \underline{R}Y$

(6) $X \subseteq Y$ implies $\overline{R}X \subseteq \overline{R}Y$
(7) $\underline{R}(X \cup Y) \supseteq \underline{R}X \cup \underline{R}Y$
(8) $\overline{R}(X \cap Y) \subseteq \overline{R}X \cap \overline{R}Y$
(9) $\underline{R}(-X) = -\overline{R}X$
(10) $\overline{R}(-X) = -\underline{R}X$
(11) $\underline{R}\underline{R}X = \overline{R}\underline{R}X = \underline{R}X$
(12) $\overline{R}\overline{R}X = \underline{R}\overline{R}X = \overline{R}X$.

After Pawlak proposed a rough set theory, many people developed its variants and extensions. For example, in 1993, Ziarko generalized Pawlak's original rough set model in Ziarko [65], developing the so-called *variable precision rough set model* (VPRS model). It can overcome the inability to model uncertain information, and is directly derived from the original model without any additional assumptions.

As the limitations of Pawlak's rough set model, Ziarko pointed out two points. One is that it cannot provide a classification with a controlled degree of uncertainty. Some level of uncertainty in the classification process gives a deeper or better understanding for data analysis.

Another is that the original model has the assumption that the universe U of data objects is known. This means that all conclusions derived from the model are applicable only to this set of objects. It is thus useful to introduce uncertain hypotheses about properties of a larger universe.

Ziarko's extended rough set model generalizes the standard set inclusion relation, capable of allowing for some degree of misclassification in the largely correct classification.

Let X and Y be non-empty subsets of a finite universe U. X is included in Y, denoted $Y \supseteq X$, if for all $e \in X$ implies $e \in Y$. Here, we introduce the measure $c(X, Y)$ of the relative degree of misclassification of the set X with respect to set Y defined as:

$$c(X, Y) = 1 - \text{card}(X \cap Y)/\text{card}(X) \text{ if card}(X) > 0 \text{ or}$$
$$c(X, Y) = 0 \text{ if card}(X) = 0$$

where card denotes set cardinality.

The quantity $c(X, Y)$ will be referred to as the relative classification error. The actual number of misclassification is given by the product $c(X, Y) * \text{card}(X)$ which is referred to as an absolute classification error.

We can define the inclusion relationship between X and Y without explicitly using a general quantifier:

$$X \subseteq Y \text{ iff } c(X, Y) = 0.$$

The *majority* requirement implies that more than 50% of X elements should be in common with Y. The *specified majority* requirement imposes an additional requirement. The number of elements of X in common with Y should be above 50% and not below a certain limit, e.g. 85%.

According to the specified majority requirement, the admissible classification error β must be within the range $0 \leq \beta < 0.5$. Then, we can define the majority inclusion relation based on this assumption.

$$X \overset{\beta}{\subseteq} Y \text{ iff } c(X, Y) \leq \beta.$$

The above definition covers the whole family of β-majority relation. However, the majority inclusion relation does not have the transitivity relation.

For the VPRS-model, we define the approximation space as a pair $A = (U, R)$, where U is a non-empty finite universe and R is an equivalence relation on U. The equivalence relation R, referred to as an indiscernibility relation, corresponds to a partitioning of the universe U into a collection of equivalence classes or elementary set $R^* = \{E_1, E_2, \ldots, E_n\}$.

Using a majority inclusion relation instead of the inclusion relation, we can obtain generalized notions of β-lower approximation (or β-positive region $\text{POSR}_\beta(X)$) of the set $U \supseteq X$:

$$\underline{R}_\beta X = \bigcup \{E \in R^* \mid X \overset{\beta}{\supseteq} E\} \text{ or, equivalently,}$$
$$\underline{R}_\beta X = \bigcup \{E \in R^* \mid c(E, X) \leq \beta\}$$

The β-upper approximation of the set $U \supseteq X$ can be also defined as follows:

$$\overline{R}_\beta X = \bigcup \{E \in R^* \mid c(E, X) < 1 - \beta\}$$

The β-boundary region of a set is given by

$$\text{BNR}_\beta X = \bigcup \{E \in R^* \mid \beta < c(E, X) < 1 - \beta\}.$$

The β-negative region of X is defined as a complement of the β-upper approximation:

$$\text{NEGR}_\beta X = \bigcup \{E \in R^* \mid c(E, X) \geq 1 - \beta\}.$$

The lower approximation of the set X can be interpreted as the collection of all those elements of U which can be classified into X with the classification error not greater than β.

The β-negative region of X is the collection of all those elements of U which can be classified into the complement of X, with the classification error not greater than β.

The β-boundary region of X consists of all those elements of U which cannot be classified either into X or into $-X$ with the classification error not greater than β.

Notice here that the law of excluded middle, i.e. $p \vee \neg p$, where $\neg p$ is the negation of p, holds in general for imprecisely specified sets.

Finally, the β-upper approximation $\overline{R}_\beta X$ of X includes all those elements of U which cannot be classified into $-X$ with the error not greater than β. If $\beta = 0$ then the original rough set model is a special case of VPRS-model.

Intuitively, with the decrease of the classification error β the size of the positive and negative regions of X will shrink, whereas the size of the boundary region will grow.

With the reduction of β fewer elementary sets will satisfy the criterion for inclusion in β-positive or β negative regions. Thus, the size of the boundary will increase.

The set $\text{BNR}_{0.5}X$ is called an *absolute boundary* of X because it is included in every other boundary region of X.

The absolute boundary is very "narrow", consisting only of those sets which have 50/50 aplite of elements among set X interior and its exterior. All other elementary sets are classified either into positive region $\underline{R}_{0.5}X$ or the negative region $\text{NEGR}_{0.5}X$.

We turn to the measure of approximation. To express the degree with which a set X can be approximately characterized by means of elementary sets of the approximation space $A = (U, R)$, we will generalize the accuracy measure introduced by Pawlak.

The β-accuracy for $0 \leq \beta < 0.5$ is defined as

$$\alpha(R, \beta, X) = \text{card}(\underline{R}_\beta X)/\text{card}(\overline{R}_\beta X).$$

The β-accuracy represents the imprecision of the approximate characterization of the set X relative to assumed classification error β.

Note that with the increase of β the cardinality of the β-upper approximation will tend downward and the size of the β-lower approximation will tend upward which leads to the conclusion that is consistent with intuition that relative accuracy may increase at the expense of a higher classification error.

The notion of discernibility of set boundaries is relative. If a large classification error is allowed then the set X can be highly discernible within assumed classification limits. When smaller values of the classification tolerance are assumed in may become more difficult to discern positive and negative regions of the set to meet the narrow tolerance limits.

The set X is said to be β-*discernible* if its β-boundary region is empty or, equivalently, if

$$\underline{R}_\beta X = \overline{R}_\beta X.$$

Any set X which is not discernible for every β is called indiscernible or absolutely rough. The set X is absolutely rough iff $\text{BNR}_{0.5}X \neq \emptyset$. Any set which is not absolutely rough will be referred to as relatively rough or weakly discernible.

For each relatively rough set X, there exists such a classification error level β that X is discernible on this level.

Let $\text{NDIS}(R, X) = \{0 \leq \beta < 0.5 \mid \text{BNR}_\beta(X) \neq \emptyset\}$. Then, $\text{NDIS}(R, X)$ is a range of all those β values for which X is indiscernible.

The least value of classification error β which makes X discernible will be referred to as discernibility threshold. The value of the threshold is equal to the least upper bound $\zeta(R, X)$ of $\text{NDIS}(X)$, i.e.,

$$\zeta(R, X) = \sup \text{NDIS}(R, X).$$

The discernibility threshold of the set X equals a minimal classification error β which can be allowed to make this set β-discernible.

As is clear from the above exposition, variable precision rough set model can be regarded as a direct generalization of the original rough set model. Consult Ziarko [65] for more details.

Yao et al. proposed *decision theoretic rough set model* (DTRS model) in Yao et al. [60], which can be seen as a quantitative generalization of Pawlak rough set model by considering the degree of inclusion of an equivalence class in a set. It determines the threshold values for deciding the three-regions in rough sets by using the Bayesian decision procedure.

The acceptance-rejection evaluation used by a DTRS model is the conditional probability $v_A(x) = Pr(A \mid [x])$, with values from the totally ordered set of $([0, 1], \leq)$. In DTRS model, the required parameters for defining probabilistic lower and upper approximations are calculated based on more familiar notions of costs (risks) through the well-known *Bayesian decision procedure*. The Bayesian decision procedure deals with making decision with minimum risk based on observed evidence.

Let $\Omega = \{w_1, \ldots, w_s\}$ be a finite set of s states, and let $\mathcal{A} = \{a_1, \ldots, a_m\}$ be a finite set of m possible actions. Let $P(w_j \mid \mathbf{x})$ be the conditional probability of an object x being in state w_j given that the object is described by $\mathbf{x}$. Let $\lambda(a_i \mid w_j)$ denote the loss, or cost, for taking action a_i when the state is w_j. For an object with description $\mathbf{x}$, suppose action a_i is taken.

Since $P(w_j \mid \mathbf{x})$ is the probability that the true state is w_j given $\mathbf{x}$, the expected loss associated with taking action a_i is given by:

$$R(a_i \mid \mathbf{x}) = \sum_{j=1}^{s} \lambda(a_i \mid w_j) P(w_j \mid \mathbf{x})$$

The quantity $R(s_i \mid \mathbf{x})$ is also called the *conditional risk*.

Given a description $\mathbf{x}$, a decision rule is a function $\tau(\mathbf{x})$ that specifies which action to take. That is, for every $\mathbf{x}$, $\tau(\mathbf{x})$ takes one of the actions, $a_1, \ldots, a_m$. The overall risk $\mathbf{R}$ is the expected loss associated with action $\tau(\mathbf{x})$, the overall risk is defined by:

$$\mathbf{R} = \sum_{\mathbf{x}} R(\tau(\mathbf{x}) \mid \mathbf{x}) P(\mathbf{x})$$

where the summation is over the set of all possible descriptions of objects. If $\tau(\mathbf{x})$ is chosen so that $R(\tau(\mathbf{x}) \mid \mathbf{x})$ is as small as possible for every $\mathbf{x}$, the overall risk $\mathbf{R}$ is minimized.

In an approximation space $apr = (U, E)$, an equivalence class $[x]$ is considered to be the description of x. The partition U/E is the set of all possible descriptions. The classification of objects according to approximation operators can be easily fitted into the Bayesian decision-theoretic framework. The set of states is given by $\Omega = \{A, A^c\}$ indicating that an element is in A and not in A, respectively.

We use the same symbol to denote both a subset A and the corresponding state. With respect to the three regions, the set of actions is given by $\mathcal{A} = \{a_1, a_2, a_3\}$ where a_1, a_2 and a_3 represent the three actions in classifying an object, deciding $POS(A)$, deciding $NEG(A)$, and deciding $BND(A)$, respectively.

Let $\lambda(a_i \mid A)$ denote the loss incurred for taking action a_i when an object belong to A, and let $\lambda(a_i \mid A^c)$ denote the loss incurred for taking the same action when the object does not belong to A.

$$R(a_1 \mid [x]) = \lambda_{11} P(A \mid [x]) + \lambda_{12} P(A^c \mid [x]),$$
$$R(a_2 \mid [x]) = \lambda_{21} P(A \mid [x]) + \lambda_{22} P(A^c \mid [x]),$$
$$R(a_3 \mid [x]) = \lambda_{31} P(A \mid [x]) + \lambda_{32} P(A^c \mid [x]),$$

where $\lambda_{i1} = \lambda(a_i \mid A)$, $\lambda_{i2} = \lambda(a_i \mid A^c)$ ($i = 1, 2, 3$). The Bayesian decision procedure leads to the following minimum-risk decision rules:

(P) If $R(a_1 \mid [x]) \leq R(a_2 \mid [x])$ and $R(a_1 \mid [x]) \leq R(a_3 \mid [x])$, decide $POS(A)$,
(N) If $R(a_2 \mid [x]) \leq R(a_1 \mid [x])$ and $R(a_2 \mid [x]) \leq R(a_3 \mid [x])$, decide $NEG(A)$,
(B) If $R(a_3 \mid [x]) \leq R(a_1 \mid [x])$ and $R(a_3 \mid [x]) \leq R(a_2 \mid [x])$, decide $BND(A)$.

Tie-breaking criteria should be added so that each element is classified into only one region. Since $P(A \mid [x]) + P(A^c \mid [x]) = 1$, we can simplify the rules to classify any object in $[x]$ based only on probabilities $P(A \mid [x])$ and the loss function λ_{ij} ($i = 1, 2, 3, j = 1, 2$).

Lin gave several examples of decision rules according to the conditions for a loss function. For example, consider a special kind of loss function with $\lambda_{11} \leq \lambda_{31} < \lambda_{21}$ and $\lambda_{22} \leq \lambda_{32} < \lambda_{12}$. That is, the loss of classifying an object x belonging to A into the positive region $POS(A)$ is less than or equal to the loss of classifying x into the boundary region $BND(A)$, and both of those losses are strictly less than the loss of classifying x into the negative region $NEG(A)$.

The reverse order of losses is used for classifying an object that does not belong to A. For this type of loss function, the minimum-risk decision rules (P)-(B) can be written as;

(P) If $P(A \mid [x]) \geq \gamma$ and $P(A \mid [x]) \geq \alpha$, decide $POS(A)$
(N) If $P(A \mid [x]) \leq \beta$ and $P(A \mid [x]) \leq \gamma$, decide $NEG(A)$
(B) If $\beta \leq P(A \mid [x]) \leq \alpha$, decide $BND(A)$

where

$$\alpha = \frac{\lambda_{12} - \lambda_{32}}{(\lambda_{31} - \lambda_{32}) - (\lambda_{11} - \lambda_{12})}$$
$$\gamma = \frac{\lambda_{12} - \lambda_{22}}{(\lambda_{21} - \lambda_{22}) - (\lambda_{11} - \lambda_{12})}$$
$$\beta = \frac{\lambda_{32} - \lambda_{22}}{(\lambda_{21} - \lambda_{22}) - (\lambda_{31} - \lambda_{32})}$$

By the assumptions, $\lambda_{11} \leq \lambda_{31} < \lambda_{21}$ and $\lambda_{22} \leq \lambda_{32} < \lambda_{12}$, it follows that $\alpha \in (0, 1]$, $\gamma \in (0, 1)$, and $\beta \in [0, 1)$.

For a loss function with $\lambda_{11} \leq \lambda_{31} < \lambda_{21}$ and $\lambda_{22} \leq \lambda_{32} < \lambda_{12}$, more results about the required parameters α, β and γ are summarized as follows:

1. If a loss function satisfies the condition $(\lambda_{12} - \lambda_{32})(\lambda_{21} - \lambda_{31}) \geq (\lambda_{31} - \lambda_{11})$ $(\lambda_{32} - \lambda_{22})$ then $\alpha \geq \gamma \geq \beta$.
2. If a loss function satisfies the condition $(\lambda_{12} - \lambda_{22})(\lambda_{31} - \lambda_{11})$, then $\alpha \geq 0.5$.
3. If a loss function satisfies the conditions $(\lambda_{12} - \lambda_{22})(\lambda_{31} - \lambda_{11})$ and $(\lambda_{12} - \lambda_{32})(\lambda_{21} - \lambda_{31}) \geq (\lambda_{31} - \lambda_{11})(\lambda_{32} - \lambda_{22})$ then $\alpha \geq 0.5$ and $\alpha \geq \beta$.
4. If a loss function satisfies the condition $(\lambda_{12} - \lambda_{32})(\lambda_{32} - \lambda_{22}) = (\lambda_{31} - \lambda_{11})$ $(\lambda_{31} - \lambda_{11})(\lambda_{21} - \lambda_{31})$, then $\beta = 1 - \alpha$.
5. If a loss function satisfies the two sets of equivalent conditions,

   (i) $(\lambda_{12} - \lambda_{32})(\lambda_{21} - \lambda_{31}) \geq (\lambda_{31} - \lambda_{11})(\lambda_{32} - \lambda_{22})$,
   $\quad (\lambda_{12} - \lambda_{32})(\lambda_{32} - \lambda_{22}) = (\lambda_{31} - \lambda_{11})(\lambda_{21} - \lambda_{31})$,
   (ii) $\lambda_{12} - \lambda_{32} \geq \lambda_{31} - \lambda_{11}$, $(\lambda_{12} - \lambda_{32})(\lambda_{32} - \lambda_{22}) = (\lambda_{31} - \lambda_{11})(\lambda_{21} - \lambda_{31})$,

   then $\alpha = 1 - \beta \geq 0.5$.

Here, the condition of Case 1 guarantees that the probabilistic lower approximation of a set is a subset of its probabilistic upper approximation. The condition of Case 2 ensures that a lower approximation of A consists of those elements whose majority equivalent elements are in A. The condition of Case 4 results in a pair of dual lower and upper approximation operators. Case 3 is a combination of Cases 1 and 2. Case 5 is the combination of Cases 1 and 4 or the combination of Cases 3 and 4.

When $\alpha > \beta$, we have $\alpha > \gamma > \beta$. After tie-breaking, we obtain the decision rules:

(P1) If $P(A \mid [x]) \geq \alpha$, decide $POS(A)$
(N1) If $P(A \mid [x]) \leq \beta$ decide $NEG(A)$
(B1) If $\beta < P(A \mid [x]) < \alpha$, decide $BND(A)$

When $\alpha = \beta$, we have $\alpha = \gamma = \beta$. In this case, we use the decision rules:

(P2) If $P(A \mid [x]) > \alpha$, decide $POS(A)$
(N2) If $P(A \mid [x]) < \alpha$ decide $NEG(A)$
(B2) If $P(A \mid [x]) = \alpha$, decide $BND(A)$

For the second set of decision rules, we use a tie-breaking criterion so that the boundary region may be non-empty.

DTRS model can offer new insights into the probabilistic approaches to rough sets. And it can generalize previous approaches to rough set models, including Pawlak rough set model and VPRS model. We will discuss DTRS model as the underlying basis for a *theory of three-way decisions* in Chap. 12.

Rough set theory can properly handle knowledge in table from. In this connection, we mention that *formal concept analysis* (FCA) due to Ganter and Wille [17] can be viewed as a similar theory for knowledge representation. The underlying basis of FCA is *concept lattice* to model relations of concept in a precise way. Obviously, rough set theory and formal concept analysis share similar idea.

FCA uses the notion of a formal concept as a mathematical formulation of the notion of a concept in Port-Royal logic. According to Port-Royal, a concept is determined by a collection of objects, called an *extent* which fall under the concept and a collection of attributes called an *intent* covered by the concepts. Concepts are ordered by a subconcept-superconcept relation which is based on inclusion relation on objects and attributes.

A *formal context* is a triplet $\langle X, Y, I \rangle$, where X and Y are non-empty set, and I is a binary relation, i.e., $I \subseteq X \times Y$. Elements x from X are called *objects*, elements y from Y are called *attributes*, $\langle x, y \rangle \in I$ indicates that x has attribute y.

For a given cross-table with n rows and m columns, a corresponding formal context $\langle X, Y, I \rangle$ consists of a set $X = \{x_1, \ldots, x_n\}$, a set $Y = \{y_1, \ldots, y_m\}$, and a relation I defined by $\langle x_i, y_j \rangle \in I$ iff the table entry corresponding to row i and column j contains $\times$.

Concept-forming operators are defined for every formal context. For a formal context $\langle X, Y, I \rangle$, operators $^{\uparrow} : 2^X \to 2^Y$ and $^{\downarrow} : 2^Y \to 2^X$ are defined for every $A \subseteq X$ and $B \subseteq Y$ by

$$A^{\uparrow} = \{y \in Y \mid \text{for each } x \in A : \langle x, y \rangle \in I\}$$
$$B^{\downarrow} = \{x \in X \mid \text{for each } y \in B : \langle x, y \rangle \in I\}$$

Formal concepts are particular clusters in cross-tables, defined by means of attribute sharing. A *formal concept* in $\langle X, Y, I \rangle$ is a pair $\langle A, B \rangle$ of $A \subseteq X$ and $B \subseteq Y$ such that $A^{\uparrow} = B$ and $B^{\downarrow} = A$.

It is noticed that $\langle A, B \rangle$ is a formal concept iff A contains just objects sharing all attributes from B and B contains just attributes shared by all objects from A. Thus, mathematically $\langle A, B \rangle$ is a formal concept iff $\langle A, B \rangle$ is a fixpoint of the pair $\langle \uparrow, \downarrow \rangle$ of the concept-forming operators.

Consider the following table:

I	y_1	y_2	y_3	y_4
x_1	$\times$	$\times$	$\times$	$\times$
x_2	$\times$		$\times$	$\times$
x_3		$\times$	$\times$	$\times$
x_4		$\times$	$\times$	$\times$
x_5	$\times$			

Here, formal concept

$$\langle A_1, B_1 \rangle = \langle \{x_1, x_2, x_3, x_4\}, \{y_3, y_4\} \rangle$$

because

$$\{x_1, x_2, x_3, x_4\}^{\uparrow} = \{y_3, y_4\}$$
$$\{y_3, y_4\}^{\downarrow} = \{x_1, x_2, x_3, x_4\}$$

Here, the following relationships hold:

$$\{x_2\}^\uparrow = \{y_1, y_3, y_4\}, \{x_2, x_3\}^\uparrow = \{y_3, y_4\}$$
$$\{x_1, x_4, x_5\}^\uparrow = \emptyset$$
$$X^\uparrow = \emptyset, \emptyset^\uparrow = Y$$
$$\{y_1\}^\downarrow = \{x_1, x_2, x_5\}, \{y_1, y_2\}^\downarrow = \{x_1\}$$
$$\{y_2, y_3\}^\downarrow = \{x_1, x_3, x_4\}, \{y_2, y_3, y_4\}^\downarrow = \{x_1, x_3, x_4\}$$
$$\emptyset^\downarrow = X, Y^\downarrow = \{x_1\}$$

Concepts are naturally ordered by a subconcept-superconcept relation. The subconcept-superconcept relation, denoted $\leq$, is based on inclusion relation on objects and attributes. For formal concepts $\langle A_1, B_1 \rangle$ and $\langle A_2, B_2 \rangle$ of $\langle A, Y, I \rangle$, $\langle A_1, B_1 \rangle \leq \langle A_2, B_2 \rangle$ iff $A_1 \subseteq A_2$ (iff $B_2 \subseteq B_1$).

In the above example, the following hold:

$$\langle A_1, B_1 \rangle = \{\{x_1, x_2, x_3, x_4\}, \{y_3, y_4\}\}$$
$$\langle A_2, B_2 \rangle = \{\{x_1, x_3, x_4\}, \{y_2, y_3, y_4\}\}$$
$$\langle A_3, B_3 \rangle = \{\{x_1, x_2\}, \{y_1, y_3, y_4\}\}$$
$$\langle A_4, B_4 \rangle = \langle\{x_1, x_2, x_5\}, \{y_1\}\rangle$$
$$\langle A_3, B_3 \rangle \leq \langle A_1, B_1 \rangle$$
$$\langle A_3, B_3 \rangle \leq \langle A_4, B_4 \rangle$$
$$\langle A_2, B_2 \rangle \leq \langle A_1, B_1 \rangle$$
$$\langle A_1, B_1 \rangle \parallel \langle A_4, B_4 \rangle \text{ (incomparable)}$$
$$\langle A_2, B_2 \rangle \parallel \langle A_3, B_3 \rangle$$
$$\langle A_2, B_2 \rangle \parallel \langle A_4, B_4 \rangle$$

We denote by $\mathcal{B}(X, Y, I)$ the collection of all formal concepts of $\langle X, Y, I \rangle$, i.e.,

$$\mathcal{B}(X, Y, I) = \{\langle A, B \rangle \in 2^X \times 2^Y \mid A^\uparrow = B, B^\downarrow = A\}.$$

$\mathcal{B}(X, Y, I)$ equipped with the subconcept-superconcept ordering $\leq$ is called a *concept lattice* of $\langle X, Y, I \rangle$. $\mathcal{B}(X, Y, I)$ represents all clusters which are hidden in data $\langle X, Y, I \rangle$. We can see that $\langle \mathcal{B}(X, Y, I), \leq \rangle$ is a lattice.

Extents and intents of concepts are defined as follows:

$\text{Ext}(X, Y, I) = \{A \in 2^X \mid \langle A, B \rangle \in \mathcal{B}(X, Y, I) \text{ for some } B\}$ (extents of concepts)
$\text{Int}(X, Y, I) = \{A \in 2^Y \mid \langle A, B \rangle \in \mathcal{B}(X, Y, I) \text{ for some } A\}$ (extents of concepts)

Formal concepts can be also defined as maximal rectangles in the cross-table: A rectangle in $\langle X, Y, I \rangle$ is a pair $\langle A, B \rangle$ such that $A \times B \subseteq I$, i.e., for each $x \in A$ and $y \in B$ we have $\langle x, y \rangle \in I$. For rectangles $\langle A_1, B_1 \rangle$ and $\langle A_2, B_2 \rangle$, put $\langle A_1, B_1 \rangle \sqsubseteq \langle A_2, B_2 \rangle$ iff $A_1 \subseteq A_2$ and $B_1 \subseteq B_2$.

We can prove that $\langle A, B \rangle$ is a formal concept of $\langle X, Y, I \rangle$ iff $\langle A, B \rangle$ is a maximal rectangle in $\langle X, Y, I \rangle$.

Consider the following table.

I	y_1	y_2	y_3	y_4
x_1	×	×	×	×
x_2	×		×	×
x_3		×	×	×
x_4		×	×	×
x_5	×			

In this table, $\langle \{x_1, x_2, x_3\}, \{y_3, y_4\} \rangle$ is a rectangle which is not maximal with respect to $\sqsubseteq$. $\langle \{x_1, x_2, x_3, x_4\}, \{y_3, y_4\} \rangle$ is a rectangle which is maximal with respect to $\sqsubseteq$. The notion of rectangle can server as a basis for geometrical reasoning in formal concept analysis.

There are two basic mathematical structures behind formal concept analysis, i.e., Galois connections and closure operators. A *Galois connection* (cf. Ore [38]) between sets X and Y is a pair $\langle f, g \rangle$ of $f : 2^X \rightarrow 2^Y$ and $g : 2^Y \rightarrow 2^X$ satisfying $A, A_1, A_2, B, B_1, B_2 \subseteq Y$:

$$A_1 \subseteq A_2 \ \Rightarrow \ f(A_2) \subseteq f(A_1)$$
$$B_1 \subseteq B_2 \ \Rightarrow \ g(B_2) \subseteq f(B_1)$$
$$A \subseteq g(f(A))$$
$$B \subseteq f(g(B)).$$

For a Galois connection $\langle f, g \rangle$ between sets X and Y, the set:

$$\text{fix}(\langle f, g \rangle) = \{\langle A, B \rangle \in 2^X \times 2^Y \mid f(A) = B, g(B) = A\}$$

is called a set of *fixpoint* of $\langle f, g \rangle$.

Here, we show a basic property of concept-forming operators. That is, for a formal context $\langle X, Y, I \rangle$, the pair $\langle \uparrow_I, \downarrow_I \rangle$ of operators induced by $\langle X, Y, I \rangle$ is a Galois connection between X and Y.

As consequence of the property, it is shown that for a Galois connection $\langle f, g \rangle$ between X and Y, $f(A) = f(g(f(A)))$ and $g(B) = g(f(g(B)))$ for any $A \subseteq X$ and $B \subseteq Y$.

Closure operators result from the concept-forming operators by their composition. If $\langle f, g \rangle$ is a Galois connection between X and Y, then $C_X = f \circ g$ is a closure operator on X and $C_Y = g \circ f$ is a closure operator on Y.

We can show that extents and intents are just the images under the concept-forming operators as follows:

$$\text{Ext}(X, Y, I) = \{B^\downarrow \mid B \subseteq Y\}$$
$$\text{Int}(X, Y, I) = \{A^\uparrow \mid A \subseteq X\}.$$

The following relationships hold for any formal context $\langle X, Y, I \rangle$:

$$\text{Ext}(X, Y, I) = \text{fix}(\uparrow\downarrow)$$
$$\text{Int}(X, Y, I) = \text{fix}(\downarrow\uparrow)$$
$$\mathcal{B}(X, Y, I) = \{\langle A, A^\uparrow \rangle \mid A \in \text{Ext}(X, Y, I)\}$$
$$\mathcal{B}(X, Y, I) = \{\langle B^\downarrow, B \rangle \mid B \in \text{Int}(X, Y, I)\}$$

The above definition of Galois connection can be simplified by the following simplified form. $\langle f, g \rangle$ is a Galois connection between X and Y iff for every $A \subseteq X$ and $B \subseteq Y$:

$A \subseteq g(B)$ iff $B \subseteq f(A)$.

Galois connections with respect to union and intersection satisfy the following properties: Let $\langle f, g \rangle$ be a Galois connection between X and Y. For $A_j \subseteq X, j \in J$ and $B_j \subseteq Y, j \in J$, we have:

$$f\left(\bigcup_{j \in J} A_j\right) = \bigcap_{j \in J} f(A_j)$$
$$g\left(\bigcup_{j \in J} B_j\right) = \bigcap_{j \in J} g(B_j)$$

Every pair of concept-forming operators forms a Galois connection, and every Galois connection is a concept-forming operator of a particular formal context.

Let $\langle f, g \rangle$ be a Galois connection between X and Y. Consider a formal context $\langle X, Y, I \rangle$ such that I is defined by

$\langle x, y \rangle \in I$ iff $y \in f(\{x\})$ or equivalently, iff $x \in g(\{y\})$

for each $x \in X$ and $y \in Y$. Then, $\langle \uparrow_I, \downarrow_I \rangle = \langle f, g \rangle$, i.e., $\langle \uparrow_I, \downarrow_I \rangle$ induced by $\langle X, Y, I \rangle$ coincide with $\langle f, g \rangle$.

We can establish representation result in the following form, i.e., $I \mapsto \langle \uparrow_I, \downarrow_I \rangle$ and $\langle \uparrow_I, \downarrow_I \rangle \mapsto I_{\langle \uparrow, \downarrow \rangle}$ are mutually inverse mappings between the set of all binary relations between X and Y and the set of all Galois connections between X and Y.

We can also see the duality relationships between extents and intents. For $\langle A_1, B_1 \rangle, \langle A_2, B_2 \rangle \in \mathcal{B}(X, Y, I)$, we have that $A_1 \subseteq A_2$ iff $B_2 \subseteq B_1$. Then, we have the following properties:

(1) $\langle \text{Ext}(X, Y, I), \subseteq \rangle$ and $\langle \text{Int}(X, Y, I), \subseteq \rangle$ are partially ordered sets.
(2) $\langle \text{Ext}(X, Y, I), \subseteq \rangle$ and $\langle \text{Int}(X, Y, I), \subseteq \rangle$ are dually isomorphic, i.e., there is a mapping $f : \text{Ext}(X, Y, I) \to \text{Int}(X, Y, I)$ satisfying $A_1 \subseteq A_2$ iff $f(A_2) \subseteq f(A_1)$.
(3) $\langle \mathcal{B}(X, Y, I), \leq \rangle$ is isomorphic to $\langle \text{Ext}(X, Y, I), \subseteq \rangle$.
(4) $\langle \mathcal{B}(X, Y, I), \leq \rangle$ is dually isomorphic to $\langle \text{Int}(X, Y, I), \subseteq \rangle$.

We can also state the property of fixpoints of closure operators. For a closure operator C on X, the partially ordered set $\langle \text{fix}(C), \subseteq \rangle$ of fixpoints of C is a complete lattice with infima and suprema given by:

$$\bigwedge_{j \in J} A_j = C\left(\bigcap_{j \in J} A_j\right)$$
$$\bigvee_{j \in J} A_j = C\left(\bigcup_{j \in J} A_j\right)$$

The following is the main result of concept lattices due to Wille.

(1) $\mathcal{B}(X, Y, I)$ is a complete lattice with infima and suprema given by

$$\bigwedge_{j\in J}\langle A_j, B_j\rangle = \langle\bigcap_{j\in J}A_j, (\bigcup_{j\in J}B_j)^{\downarrow\uparrow}\rangle,$$

$$\bigvee_{j\in J}\langle A_j, B_j\rangle = \langle(\bigcup_{j\in J}A_j)^{\uparrow\downarrow}, \bigcap_{j\in J}B_j\rangle.$$

(2) Moreover, an arbitrary complete lattice $\mathbf{V} = (V, \leq)$ is isomorphic to $\mathcal{B}(X, Y, I)$ iff there are mapping $\gamma : X \to V, \mu : Y \to X$ such that

    (i) $\gamma(X)$ is $\bigvee$-dense in V, $\mu(Y)$ is $\bigwedge$-dense in V

    (ii) $\gamma(x) \leq \mu(y)$ iff $\langle x, y\rangle \in I$.

In formal concept analysis, we can clarify and reduce formal concepts by removing some of objects or attributes in a formal context. A formal context $\langle X, Y, I\rangle$ is called *clarified* if the corresponding table neither contain identical rows nor identical columns. Namely, if $\langle X, Y, I\rangle$ is clarified then:

$\{x_1\}^{\uparrow} = \{x_2\}^{\uparrow}$ implies $x_1 = x_2$ for every $x_1, x_2 \in X$,
$\{y_1\}^{\downarrow} = \{y_2\}^{\downarrow}$ implies $y_1 = y_2$ for every $y_1, y_2 \in Y$.

Clarification can be performed by removing identical rows and columns. If $\langle X_1, Y_1, I_1\rangle$ is a clarified context resulting from $\langle X_2, Y_2, I_2\rangle$ by clarification, then $\mathcal{B}(X_1, Y_1, I_1)$ is isomorphic to $\mathcal{B}(X_2, Y_2, I_2)$.

For a formal context $\langle X, Y, I\rangle$, an attribute $y \in Y$ is called *reducible* iff there is $Y' \subset Y$ with $y \notin Y'$ such that

$$\{y\}^{\downarrow} = \bigcap_{z\in Y'}\{z\}^{\downarrow}$$

i.e., the column corresponding to y is the intersection of columns corresponding to z's from Y'.

An object $x \in X$ is called *reducible* iff there is $X' \subset X$ with $x \notin X'$ such that

$$\{x\}^{\uparrow} = \bigcap_{z\in X'}\{z\}^{\uparrow}$$

i.e., the row corresponding to x is the intersection of columns corresponding to z's from X'.

Let $y \in Y$ be reducible in $\langle X, Y, I\rangle$. Then, $\mathcal{B}(X, Y - \{y\}, J)$ is isomorphic to $\mathcal{B}(X, Y, I)$, where $J = I \cap (X \times (Y - \{y\}))$ is the restriction of I to $X \times Y - \{y\}$, i.e., $\langle X, Y - \{y\}, J\rangle$ results by removing y from $\langle X, Y, I\rangle$.

$\langle X, Y, I\rangle$ is *row reducible* if no object $x \in X$ is reducible; it is *column reducible* if no attribute $y \in Y$ is reducible; it is *reduced* if it is both row reduced and column reduced.

Arrow relations can find which objects and attributes are reducible. For $\langle X, Y, I\rangle$, we define relations $\nearrow$, $\swarrow$, $\updownarrow$ between X and Y:

$x \swarrow y$ iff $\langle x, y\rangle \notin I$ and if $\{x\}^{\uparrow} \subset \{x_1\}^{\uparrow}$ then $\langle x_1, y\rangle \in I$
$x \nearrow y$ iff $\langle x, y\rangle \notin I$ and if $\{x\}^{\downarrow} \subset \{x_1\}^{\downarrow}$ then $\langle x_1, y\rangle \in I$
$x \updownarrow y$ iff $x \swarrow y$ and $x \nearrow y$.

Thus, if $\langle x, y \rangle \in I$, then none of the above three relations occurs. Consequently, the arrow relations can be entered in the table of $\langle X, Y, I \rangle$. There is the following connections between arrow relations and reducibility.

$(\{x\}^{\uparrow\downarrow}, \{x\}^{\uparrow})$ is $\bigvee$-irreducible iff there is $y \in Y$ such that $x \swarrow y$
$(\{y\}^{\downarrow}, \{y\}^{\downarrow\uparrow})$ is $\bigwedge$-irreducible iff there is $x \in X$ such that $x \nearrow y$.

Formal concept analysis can also deal with *attribute implication* concerning dependencies of data. Let Y be a non-empty set of attributes.

An attribute implication over Y is an expression

$$A \Rightarrow B$$

where $A, B \subseteq Y$.

An attribute implication $A \Rightarrow B$ over Y is *true* (valid) in a set $M \subset Y$ iff $A \subseteq M$ implies $B \subseteq M$. We write $\| A \Rightarrow B \|_M = 1$ (0) if $A \Rightarrow B$ is true (false) in M.

Let M be a set of attributes of some object x, $\| A \Rightarrow B \|_M = 1$ says "if x has all attributes from A then x has all attributes from B", because "if x has all attributes from C" is equivalent to $C \subseteq M$.

It is possible to extend the validity of $A \Rightarrow B$ to collections $\mathcal{M}$ of M's (collections of subsets of attributes), i.e., define validity of $A \Rightarrow B$ in $\mathcal{M} \subseteq 2^Y$.

An attribute implication $A \Rightarrow B$ over Y is true (valid) in $\mathcal{M}$ if $A \Rightarrow B$ is true in each $M \in \mathcal{M}$. An attribute implication $A \Rightarrow B$ over Y is true (valid) in a table (formal context) $\langle X, Y, I \rangle$ iff $A \Rightarrow B$ is true in $\mathcal{M} = \{\{x\}^{\uparrow} \mid x \in X\}$.

We define semantic consequence (entailment). An attribute implication $A \Rightarrow B$ follows semantically from a theory T, denoted $T \models A \Rightarrow B$ iff $A \Rightarrow B$ is true in every model M of T.

The system for reasoning about attribute implications consists of the following deduction rules:

(Ax) infer $A \cup B \Rightarrow A$,
(Cut) from $A \Rightarrow B$ and $B \cup C \Rightarrow D$ infer $A \cup C \Rightarrow D$.

Note that the above deduction rules are due to Armstrong's work on functional dependencies in databases; see Armstrong [5].

A *proof* of $A \Rightarrow B$ from a set T of attribute implications is a sequence $A_1 \Rightarrow B_1, \ldots, A_n \Rightarrow B_n$ of attribute implications satisfying:

(1) $A_n \Rightarrow B_n$ is just $A \Rightarrow B$,
(2) for every $i = 1, 2, \ldots, n$:

either $A_i \Rightarrow B_i \in T$ (assumption)
or $A_i \Rightarrow B_i$ results by using (Ax) or (Cut) from preceding attribute implications $A_j \Rightarrow B_j$'s (deduction)

If we have a proof of $A \Rightarrow B$ from T, then we write $T \vdash A \Rightarrow B$. We have the following derivable rules:

(Ref) infer $A \Rightarrow A$,
(Wea) from $A \Rightarrow B$, infer $A \cup C \Rightarrow B$,
(Add) from $A \Rightarrow B$ and $A \Rightarrow C$, infer $A \Rightarrow B \cup C$,
(Pro) from $A \Rightarrow B \cup C$, infer $A \Rightarrow B$,
(Tra) from $A \Rightarrow B$ and $B \Rightarrow C$, infer $A \Rightarrow C$,

for every $A, B, C, D \subseteq Y$.

We can show that (Ax) and (Cut) are sound. It is also possible to prove soundness of above derived rules.

We can define two notions of consequence, i.e, *semantic consequence* and *syntactic consequence*:

Semantic: $T \models A \Rightarrow B$ ($A \Rightarrow B$ semantically follows from T)
Syntactic: $T \vdash A \Rightarrow B$ ($A \Rightarrow B$ syntactically follows from T)

Semantic closure of T is the set

$$sem(T) = \{A \Rightarrow B \mid T \models A \Rightarrow B\}$$

of all attribute implications which semantically follows from T.

Syntactic closure of T is the set

$$syn(T) = \{A \Rightarrow B \mid T \vdash A \Rightarrow B\}$$

of all attribute implications which syntactically follows from T.

T is *semantically closed* if $T = sem(T)$. T is *syntactically closed* if $T = syn(T)$. Note that $sem(T)$ is the least set of attribute implications which is semantically closed containing T and that $synT$) is the least set of attribute implications which is syntactically closed containing T.

It can be proved that T is syntactically closed iff for any $A, B, C, D \subseteq Y$

(1) $A \cup B \Rightarrow B \in T$,
(2) if $A \Rightarrow B \in T$ and $B \cup C \Rightarrow D \in T$ implies $A \cup C \Rightarrow D \in T$.

Then, if T is semantically closed, then T is syntactically closed. It can also be proved that if T is syntactically closed, then T is semantically closed. Consequently, soundness and completeness follow:

$T \vdash A \Rightarrow B$ iff $T \models A \Rightarrow B$.

We turn to models of attribute implications. For a set T of attribute implications, denote

$$Mod(T) = \{M \subseteq Y \mid \| A \Rightarrow B \|_M = 1 \text{ for every } A \Rightarrow B \in T\}$$

That is, $Mod(T)$ is the set of all models of T.

A closure system in a set of Y is any system S of subsets of Y which contains Y and is closed under arbitrary intersections. That is, $Y \in S$ and $\bigcap \mathcal{R} \in S$ for every $\mathcal{R} \subseteq S$ (intersection of every subsystem $\mathcal{R}$ of S belongs to S.

There is a one-to-one relationship between closure systems in Y and closure operators in Y. Namely, for a closure operator C in Y, $S_C = \{A \in 2^X \mid A = C(A)\} = \text{fix}(C)$ is a closure system in Y.

Given a closure system in Y, we set

$$C_S(A) = \bigcap \{B \in \mathcal{S} \mid A \subseteq B\}$$

for any $A \subseteq X$, C_S is a closure operator on Y. This is a one-to-one relationship, i.e., $C = C_{S_C}$ and $\mathcal{S} = \mathcal{S}_{C_S}$.

It is shown that for a set of T of attribute implications, $Mod(T)$ is a closure system in Y. Since Mod(T) is a closure system, we can consider the corresponding closure operator $C_{\mathrm{Mod}(T)}$, i.e., the fixpoints of $C_{\mathrm{Mod}(T)}$ are just models of T.

Therefore, for every $A \subseteq Y$, there exists the least model of $Mod(T)$ which contains A, namely such least models is just $C_{\mathrm{Mod}(T)}(A)$.

We can test entailment via least models follows. For any $A \Rightarrow B$ and any T, we have:

$$T \models A \Rightarrow B \text{ iff } \| A \Rightarrow B \|_{C_{\mathrm{Mod}(T)}(A)} = 1.$$

It follows that the deductive system for attribute implications is sound and complete. And it can serve as a basis for reasoning about dependencies.

FCA can be regarded as an interesting tool for data analysis. One of the advantages is that it has a mathematical foundation based on concept lattice with reasoning mechanisms based on attribute implications. Another is that FCA can visualize data in a table. Thus, we can find some similarities with rough set theory. However, since FCA uses classical (two-valued) basis, it would be difficult to extend or modify the basic formulation.

2.2 Algebras, Logics and Rough Sets

There are several approaches to characterize the concept of rough set. They are roughly divided into algebraic and logical ones. In addition, some logics for rough sets have been studied in the literature.

An *algebra* is a mathematical structure with several operations. The first algebraic approach to rough set theory is due to Iwinski [23] in 1987.

A suitable algebra for rough sets is a *double Stone algebra* (DSA). Below we review several algebraic approaches. We here assume basics of algebras.

Definition 2.1 (*Double Stone Algebra*) A double Stone algebra $\langle L, +, \cdot, *, {}^+, 0, 1 \rangle$ is an algebra of type $(2, 2, 1, 1, 0, 0)$ such that:

(1) $\langle L, +, \cdot, 0, 1 \rangle$ is a bounded distributive lattice,
(2) x^* is the pseudocomplement of x, i.e., $y \leq x^* \Leftrightarrow y \cdot x = 0$,
(3) x^+ is the dual pseudocomplement of x, i.e., $y \geq x^+ \Leftrightarrow y + x = 1$.
(4) $x^* + x^{**} = 1$, $x^+ \cdot x^{++} = 0$.

Note that conditions (2) and (3) can be described as the following equations:

$$x \cdot (x \cdot y)^* = x \cdot y^*,$$
$$x + (x + y)^+ = x + y^+,$$
$$x \cdot 0^* = x,$$

$$x + 1^+ = x,$$
$$0^{**} = 0,$$
$$1^{++} = 1.$$

DSA is called *regular* if it additionally satisfies:

$$x \cdot x^+ \leq y + y^*$$

which is equivalent to $x^+ = y^+$ and $x^* = y^*$ imply $x = y$.

The *center* $B(L) = \{x^* \mid x \in L\}$ of L is a subalgebra of L and a Boolean algebra, in which $*$ and $^+$ coincide with the Boolean complement $-$.

Lemma 2.1 states a construction of regular double Stone algebras.

Lemma 2.1 *Let $\langle B, +, \cdot, -, 0, 1 \rangle$ be a Boolean algebra and F be a not necessarily proper filter on B. Set*

$$\langle a, b \rangle^* = \langle -b, -b \rangle,$$
$$\langle a, b \rangle^+ = \langle -a, -a \rangle.$$

Furthermore $B(L) \simeq B$ as Boolean algebras, and $D(L) \simeq F$ as lattices. Note that

$$B(L) = \{\langle a, a \rangle \mid a \in B\}, \ D(L) = \{\langle a, 1 \rangle \mid a \in F\}.$$

Conversely, if M is a regular double Stone algebra, $B = B(M)$, $F = D(M)^{++}$ then the mapping which assigns to each $x \in M$ the pair $\langle x^{++}, x^{**} \rangle$ is an isomorphism M and $\langle B, F \rangle$.

From Lemma 2.1, each element x of a regular double Stone algebra is uniquely described by the greatest Boolean element below x and the smallest Boolean element above x.

Now, suppose $\langle U, \theta \rangle$ is an approximation space. Then, the classes of θ can be regarded as atoms of a complete subalgebra of the Boolean algebra $Sb(U)$. Conversely, any atomic complete subalgebra B of $Sb(U)$ yields an equivalence relation θ on U, where this correspondence is a bijection. The elements of B are $\emptyset$ and the unions of its associated equivalence relations.

If $a \in B$, then for every $X \subseteq U$ we have that if $a \in \overline{\theta}X$, then $a \in X$ (by Proposition 2.2(1)), and the rough sets of the corresponding approximation space are the elements of the regular double Stone algebra $\langle B, F \rangle$, where F is the filter of B which is generated by the union of singleton element of B. If θ is the identity on U, then $F = \{U\}$.

Other algebraic approaches can be also found in the literature. For instance, Pomykala and Pomykala [49] showed that the collection $\mathcal{B}_\theta(U)$ of rough sets of (U, θ) can be made into a Stone algebra $(\mathcal{B}_\theta(U), +, \cdot, *, (\emptyset, \emptyset), (U, U))$ by defining.

$$(\underline{X}, \overline{X}) + (\underline{Y}, \overline{Y}) = (\underline{X} \cup \underline{Y}, \overline{X} \cup \overline{Y})$$
$$(\underline{X}, \overline{X}) \cdot (\underline{Y}, \overline{Y}) = (\underline{X} \cap \underline{Y}, \overline{X} \cap \overline{Y})$$
$$(\underline{X}, \overline{X})^* = (-\overline{X}, -\overline{X})$$

where for $X \subseteq U$, the complement of Z in U is denoted by $-Z$.

There are other characterization of rough sets. As an algebraic approach, we can mention Nelson algebras and three-valued Łukasiewicz algebras. We now survey Nelson algebras for rough sets due to Pagliani [24, 44].

Nelson algebras give an algebraic semantics for *constructive logic with strong negation* of Nelson [36]; see Rasiowa [51]. In 1996, Pagliani studied relations of Nelson algebras and rough sets in [44]. Recently, in 2013, Järvinen et al. proved an algebraic completeness for constructive logic with strong negation by using finite rough set-based Nelson algebras determined by quasiorders in [24].

A *Kleene algebra* is a structure $(A, \vee, \wedge, \sim, 0, 1)$ such that A is a 0, 1-bounded distributed lattice and for all $a, b \in A$:

(K1) $\sim\sim a = a$
(K2) $a \leq b$ iff $\sim b \leq \sim a$
(K3) $a \wedge \sim a \leq b \vee \sim b$.

A *Nelson algebra* $(A, \vee, \wedge, \rightarrow, \sim, 0, 1)$ is a Kleene algebra $(A, \vee, \wedge, \sim, 0, 1)$ satisfying the following:

(N1) $a \wedge c \leq \sim a \vee b$ iff $c \leq a \rightarrow b$
(N2) $(a \wedge b) \rightarrow c = a \rightarrow (b \rightarrow c)$.

In Nelson algebras, an operation $\neg$ can be defined as $\neg a = a \rightarrow 0$. The operation $\rightarrow$ is called the *weak relative pseudocomplementation*, $\sim$ the strong negation, and $\neg$ the weak negation, respectively.

A Nelson algebra is *semi-simple* if $a \vee \neg a = 1$ for all $a \in A$. It is well known that semi-simple Nelson algebras coincide with three-valued Łukasiewicz algebras (cf. Iturrioz [22]) and regular double Stone algebras.

An element a^* in a lattice L with 0 is called a *pseudocomplement* of $a \in L$, if $a \wedge x = 0 \Leftrightarrow x \leq a^*$ for all $x \in L$. If a pseudocomplement of a exists, then it is unique.

A *Heyting algebra* H is a lattice with 0 such that for all $a, b \in H$, there is a greatest element x of H with $a \wedge x \leq b$. This element is called the *relative pseudocomplement* of a with respect to b, denoted $a \Rightarrow b$. A Heyting algebra is represented as $(H, \vee, \wedge, \Rightarrow, 0, 1)$, where the pseudocomplement of a is $a \Rightarrow 0$.

Let Θ be a Boolean congruence on a Heyting algebra H. Sendlewski [52] showed the construction of Nelson algebras from the pairs of Heyting algebras. In fact, the set of pairs

$$N_\Theta(H) = \{(a, b) \in H \times H \mid a \wedge b = 0 \text{ and } a \vee b \Theta 1\}$$

can be seen as a Nelson algebra, if we add the operations:

$(a, b) \vee (c, d) = (a \vee c, b \wedge d)$
$(a, b) \wedge (c, d) = (a \wedge c, b \vee d)$
$(a, b) \rightarrow (c, d) = (a \Rightarrow c, a \wedge d)$
$\sim (a, b) = (b, a)$

Note that $(0,1)$ is the 0-element and $(1,0)$ is the 1-element. In the right-hand side of the above equations, the operations are those of the Heyting algebra H. Sendlewski's construction can give an intuitive meaning of Vakarelov's construction for Nelson algebras in Vakarelov [56].

We now move to Pagliani's constructions. Let U be a set and E be an equivalence relation. Approximations are then defined in terms of an *indiscernibility space*, that is, a relational structure (U, E) such that E is an equivalence relation on U.

For a subset of X of U, we define the lower approximation X_E of X which consists of all elements whose E-class is included in X and the upper approximation X^E of X which is the set of the elements whose E-class has non-empty intersection with X.

Therefore, X_E can be viewed as the set of elements which *certainly* belong to X and X^E is the set of objects that *possibly* are in X, when elements are observed through the knowledge synthesized by E.

In this setting, we can use arbitrary binary relations instead of equivalence relations. For this purpose, we introduce approximations $(\cdot)_R$ and $(\cdot)^R$, where R is reflexive. They can be regarded as "real" lower and upper approximation operators.

Let R be a reflexive relation on U and $X \subseteq U$. The set $R(X) = \{y \in U \mid xRy \text{ for some } x \in X\}$ is the *R-neighbourhood* of X. If $X = \{a\}$, then we write $R(a)$ instead of $R(\{a\})$.

The approximations are defined as $X_R = \{x \in U \mid R(x) \subseteq X\}$ and $X^R = \{x \in U \mid R(X) \cap \neq \emptyset\}$. A set $X \subseteq U$ is called *R-closed* if $R(X) = X$ and an element $x \in U$ is *R-closed*, if its singleton set $\{x\}$ is R-closed. The set of R-closed points is denoted by S.

The rough set of X is the equivalence class of all $Y \subseteq U$ such that $Y_E = X_E$ and $Y^E = X^E$. Since each rough set is uniquely determined by the approximation pair, it is possible to represent the rough set of X as (X_E, X^E) or $(X_E, -X^E)$. We call the former *increasing representation* and the latter *disjoint representation*.

These representations induce the sets:

$$IRS_E(U) = \{(X_E, X^E) \mid X \subseteq U\}$$

and

$$DRS_E(U) = \{(X_E, -X^E) \mid X \subseteq U\},$$

respectively. The set $IRS_E(U)$ can be ordered componentwise

$$(X_E, X^E) \leq (Y_E, Y^E) \Leftrightarrow X_E \subseteq Y_E \text{ and } X^E \subseteq Y^E,$$

and $DRS_E(U)$ is ordered by reversing the order for the second components of the pairs:

$$(X_E. - X^E) \leq (Y_E, -Y^E) \Leftrightarrow X_E \subseteq Y_E \text{ and } -X^E \supseteq -Y^E$$
$$\Leftrightarrow X_E \subseteq Y_E \text{ and } X^E \subseteq Y^E$$

Therefore, $IRS_E(U)$ and $DRS_E(U)$ are order-isomorphic, and they form completely distributive lattices.

Because it is known that on a Boolean lattice each lattice-congruence is such that the quotient lattice is a Boolean lattice, also the congruence $\cong_S$ on $\mathcal{B}_E(U)$, defined by $X \cong_S Y$, if $X \cap S = Y \cap S$, is Boolean when $\mathcal{B}_E(U)$ is interpreted as a Heyting algebra.

Pagliani [44] showed that the disjoint representation of rough sets can be characterized as

$$DRS_E(U) = \{(A, B) \in \mathcal{B}_E(U)^2 \mid A \cap B = \emptyset \text{ and } A \cup B \cong_S U\}$$

Thus, $DRS_E(U)$ coincides with the Nelson lattice $N_{\cong_S}(\mathcal{B}_E(U))$. Since $\mathcal{B}_E(U)$ is a Boolean lattice, $N_{\cong_S}(\mathcal{B}_E(U))$ is a semi-simple Nelson algebra. Consequently, we obtain that rough sets defined by equivalences determined also regular double Stone algebras and three-valued Lukasiewicz algebras.

Järvinen et al. [24] further extended Pagliani's results for the so-called *effective lattice*, which expands Nelson algebras with a modal operator capable of expressing the classical truth. The feature is of special importance for the problems in computer science, AI, and natural language semantics.

2.3 Modal Logic and Rough Sets

It has been argued that modal logic offers foundations for rough set theory. In this line, Yao and Lin [61] provided a systematic study on the generalization of the modal logic approach. Consult Liau [27] for a comprehensive survey.

Let $apr = \langle U, R \rangle$ be an approximation space. Given an arbitrary set $A \subseteq U$, it may be impossible to describe A precisely using the equivalence classes of R. In this case, one may characterize A by a pair of lower and upper approximations:

$$\underline{apr}(A) = \bigcup_{[x]_R \subseteq A} [x]_R,$$
$$\overline{apr}(A) = \bigcup_{[x]_R \cap A \neq \emptyset} [x]_R,$$

where $[x]_R = \{y \mid xRy\}$ is the equivalence class containing x. The pair $\langle \underline{apr}(A), \overline{apr}(A) \rangle$ is the rough set with respect to A.

Since the lower approximation $\underline{apr}(A)$ is the union of all the elementary sets which are subsets of A and the $\overline{apr}(A)$ is the union of all the elementary sets which have a non-empty intersection with A, they can be described as follows:

$$\underline{apr}(A) = \{x \mid [x]_R \subseteq A\}$$
$$= \{x \in U \mid \text{for all } y \in U, xRy \text{ implies } y \in A\}$$
$$\overline{apr}(A) = \{x \mid [x]_R \cap A = \emptyset\}$$
$$= \{x \in U \mid \text{there exists a } y \in U \text{ such that } xRy \text{ and } y \in A\}$$

These interpretations are closely related to those of the necessity and possibility operators in modal logic.

Here, we list some properties of $\underline{apr}$ and $\overline{apr}$. For any subsets $A, B \subseteq U$, we have the following about $\underline{apr}$:

(AL1) $\underline{apr}(A) = \sim \overline{apr}(\sim A)$

(AL2) $\underline{apr}(U) = U$

(AL3) $\underline{apr}(A \cap B) = \underline{apr}(A) \cap \underline{apr}(B)$

(AL4) $\underline{apr}(A \cup B) \supseteq \underline{apr}(A) \cup \underline{apr}(B)$

(AL5) $A \subseteq B \Rightarrow \underline{apr}(A) \subseteq \underline{apr}(B)$

(AL6) $\underline{apr}(\emptyset) = \emptyset$

(AL7) $\underline{apr}(A) \subseteq A$

(AL8) $A \subseteq \underline{apr}(\overline{apr}(A))$

(AL9) $\underline{apr}(\overline{A}) \subseteq \underline{apr}(\underline{apr}(A))$

(AL10) $\overline{apr}(A) \subseteq \underline{apr}(\overline{apr}(A))$

and the following about $\overline{apr}$:

(AU1) $\overline{apr}(A) = \sim \underline{apr}(\sim A)$

(AU2) $\overline{apr}(\emptyset) = \emptyset$

(AU3) $\overline{apr}(A \cup B) = \overline{apr}(A) \cup \overline{apr}(B)$

(AU4) $\overline{apr}(A \cap B) \subseteq \overline{apr}(A) \cap \overline{apr}(B)$

(AU5) $A \subseteq B \Rightarrow \overline{apr}(A) \subseteq \overline{apr}(B)$

(AU6) $\overline{apr}(U) = U$

(AU7) $A \subseteq \overline{apr}(A)$

(AU8) $\overline{apr}(\underline{apr}(A)) \subseteq A$

(AU9) $\overline{apr}(\overline{apr}(A)) \subseteq \overline{apr}(A)$

(AU10) $\overline{apr}(\underline{Apr}(A)) \subseteq \underline{apr}(A)$

where $\sim A = U - A$ denotes the set complement of A. Moreover, lower and upper approximations obey properties:

(K) $\underline{apr}(\sim A \cup B) \subseteq \sim \underline{apr}(A) \cup \underline{apr}(B)$

(ALU) $\underline{apr}(A) \subseteq \overline{apr}(A)$

Properties (AL1) and (AU1) state that two approximation operators are dual operators. In fact, properties with the same number may be regarded as dual properties.

However, these properties are not independent. For example, property (AL3) implies property (AL4). Properties (AL9), (AL10), (AU9) and (AU10) are expressed in terms of set inclusion.

Rough sets described above are called *Pawlak rough sets*. They are constructed from an equivalence relation. The pair of lower and upper approximations may be interpreted as two operators $\underline{apr}$ and $\overline{apr}$ of U.

Pawlak rough set model can be interpreted by the notions of topological space and topological Boolean algebra, where the operators $\underline{apr}$ and $\overline{apr}$ can be used together with the usual set-theoretic operators $\sim$, $\cap$ and $\cup$.

Since the modal logic S5 can be understood algebraically by topological Boolean algebra, it is natural to expect that there is a connection of Pawlak rough set and modal logic S5.

Now, we describe the presentation of modal logic. Let Φ be a non-empty set of propositions, which is generated by a finite set of logical symbols $\wedge$, $\vee$, $\neg$, $\rightarrow$, modal operators $\Box$, $\Diamond$, propositional constants $\top$, $\bot$, and infinitely enumerable set

$P = \{\phi, \psi, \ldots\}$ of propositional variables. $\Box\phi$ states that ϕ is necessary and $\Diamond\phi$ states that ϕ is possible.

Let W be a non-empty set of possible worlds and R be a binary relation called accessibility relation on W. The pair (W, R) is called a *Kripke frame*. An interpretation in (W, R) is a function $v : W \times P \to \{true, false\}$, which assigns a truth value for each propositional variable with respect to each particular world w. By $v(w, a) = true$, we mean that the proposition a is true in the interpretation v in the world, written $w \models_v a$.

We extend v for all propositions in the usual way. We define $v^* : W \times \Phi \to \{true, false\}$ as follows:

(m0) for $a \in P$, $w \models_{v^*} a$ iff $w \models_v a$
(m1) $w \not\models_{v^*} \bot$, $w \models_{v^*} \top$
(m2) $w \models_{v^*} \phi \wedge \psi$ iff $w \models_{v^*} \phi$ and $w \models_{v^*} \psi$
(m3) $w \models_{v^*} \phi \vee \psi$ iff $w \models_{v^*} \phi$ or $w \models_{v^*} \psi$
(m4) $w \models_{v^*} \phi \to \psi$ iff $w \not\models_{v^*} \phi$ or $w \models_{v^*} \psi$
(m5) $w \models_{v^*} \neg\phi \vee \psi$ iff $w \not\models_{v^*} \phi$ or $w \models_{v^*} \psi$
(m6) $w \models_{v^*} \Box\phi$ iff $\forall w' \in W(wRw' \Rightarrow w' \models_{v^*} \phi)$
(m7) $w \models_{v^*} \Diamond\phi$ iff $\exists w' \in W(wRw'$ and $w' \models_{v^*} \phi)$

Here, $\not\models$ means that not $\models$. For simplicity, we will write $w \models_v \phi$, and we will drop v when it is clear from context. The necessity and possibility are dual in the sense that the following hold:

$$\Box\phi =_{def} \neg\Diamond\neg\phi,$$
$$\Diamond\phi =_{def} \neg\Box\neg\phi.$$

If the accessibility relation R is an equivalence relation, then the modal logic is S5.

We can characterize a proposition by the set of possible worlds in which it is true using a valuation function v. Based on the idea, we can define a mapping $t : \Phi \to 2^W$ as follows:

$$t(\phi) = \{w \in W : w \models \phi\}$$

Here, we call $t(\phi)$ the *truth set* of the proposition ϕ. In the truth set representation, we have:

(s1) $t(\bot) = \emptyset$, $t(\top) = W$
(s2) $t(\phi \wedge \psi) = t(\phi) \cap t(\psi)$
(s3) $t(\phi \vee \psi) = t(\phi) \cap t(\psi)$
(s4) $t(\phi \to \psi) = {\sim} t(\phi) \cup t(\psi)$
(s5) $t(\neg\phi) = {\sim} t(\phi)$
(s6) $t(\Box\phi) = \underline{apr}(t(\phi))$
(s7) $t(\Diamond\phi) = \overline{apr}(t(\phi))$

The last two properties can be derived as follows:

$$t(\Box\phi) = \{w \in W : w \models \Box\phi\}$$
$$= \{w \in W : \forall w'(wRw' \Rightarrow w' \models \phi)\}$$
$$= \{w \in W : \forall w'(wRw' \Rightarrow w' \in t(\phi))\}$$
$$= \underline{apr}(t(\phi))$$
$$t(\Diamond\phi) = \{w \in W : w \models \Diamond\phi\}$$
$$= \{w \in W : \exists w'(wRw' \text{ and } w' \models \phi)\}$$
$$= \{w \in W : \exists w'(wRw' \text{ and } w' \in t(\phi))\}$$
$$= \overline{apr}(t(\phi))$$

In the truth set interpretation above, approximation operators in Pawlak rough set model are related to modal operators in S5. There are many modal systems according to the properties on the accessibility relation in Kripke models.

Following Chellas's naming, axioms of modal logic corresponding to (AK), (ALU), (AL7)–(AL10) are given by:

(K) $\Box(\phi \rightarrow \psi) \rightarrow (\Box\phi \rightarrow \Box\psi)$
(D) $\Box\phi \rightarrow \Diamond\phi$
(T) $\Box\phi \rightarrow \phi$
(B) $\phi \rightarrow \Box\Diamond\phi$
(4) $\Box\phi \rightarrow \Box\Box\phi$
(5) $\Diamond\phi \rightarrow \Box\Diamond\phi$.

Pawlak rough set model can be characterized by modal logic S5. In fact, the standard logical operators are interpreted by usual set-theoretic operators and modal operators by rough set operators.

Since different systems of modal logic can be obtained by using various types of accessibility relations. It is thus possible to construct different rough set models by means of Kripke models for modal logic.

Yao and Lin worked out the idea by generalizing Pawlak's rough set models to provide different rough set models depending on the accessibility relation in Kripke models in Yao and Lin [61]. Their work seems to be of special interest of foundations for rough set theory. We simply review their work here.

Given a binary relation R and two elements $x, y \in U$, if xRy then we say that y is R-related to x. A binary relation can be represented by a mapping $r : U \rightarrow 2^U$:

$$r(x) = \{y \in U \mid xRy\}$$

Here, $r(x)$ consists of all R-related elements of x. We then define two unary set-theoretic operators $\underline{apr}$ and $\overline{apr}$:

$$\underline{apr}(A) = \{x : r(x) \subseteq A\}$$
$$= \{x \in U \mid \forall y \in U(xRy \Rightarrow y \in A)\}$$
$$\overline{apr}(A) = \{x : r(x) \cap A \neq \emptyset\}$$
$$= \{x \in U \mid \exists y \in U(xRy \text{ and } y \in A)\}$$

The set $apr(A)$ consists of those elements where R-related elements are all in A, and $\overline{apr}(A)$ consists of those elements such that at least one of whose R-related elements is in A.

The pair $(apr(A), \overline{apr}(A))$ denotes the *generalized rough set* of A induced by R. Operators $apr, \overline{apr} : 2^U \to 2^U$ are referred to as the generalized rough set operators. The induced system $(2^U, \cap, \cup, \sim, apr, \overline{apr})$ is called an *algebraic rough set model*.

Equations (s6) and (s7) hold for generalized rough set operators. When R is an equivalence relation, generalized rough set operators reduce to the operators in Pawlak rough set model.

We are now ready to present the classification of algebraic rough set models. First of all, we do not expect that generalized rough set operators satisfy all the properties in Pawlak rough set models. But properties (AL1)–(AL5) and (AU1)–(AU5) hold in any rough set model.

In modal logic, properties corresponding to (AK) and (AL6)–(AL10) yield different systems, and we can use these properties for various rough set models. Here, we relabel some properties used in modal logic:

(K) $apr(\sim A \cup B) \subseteq \sim apr(A) \cup apr(B)$
(D) $apr(A) \subseteq \overline{apr}(A)$
(T) $apr(A) \subseteq A$
(B) $A \subseteq apr(\overline{apr}(A))$
(4) $apr(A) \subseteq apr(apr(A))$
(5) $\overline{apr}(A) \subseteq apr(\overline{apr}(A))$

We need certain conditions on the binary relation R to construct a rough set model so that the above properties hold.

A relation R is serial if for all $x \in U$ there exists a $y \in U$ such that xRy. A relation is reflexive if for all $x \in U$, xRx holds. A relation is symmetric if for all $x, y \in U$, xRy implies yRx. A relation is transitive if for all $x, y, z \in U$, xRy and yRz imply xRz. A relation is Euclidean if for all $x, y, z \in U$, xRy and xRz imply yRz. The corresponding approximation operators are as follows:

(serial)   forall $x \in U$, $r(x) \neq \emptyset$
(reflexive)   forall $x \in U$, $x \in r(x)$
(symmetric) forall $x, y \in U$, if $x \in r(y)$, then $y \in r(x)$
(transitive)   forall $x, y \in U$, if $y \in r(x)$, then $r(y) \subseteq r(x)$
(Euclidean) forall $x, y \in U$, if $y \in r(x)$, then $r(x) \subseteq r(y)$

Theorem 2.1 says rough set models with specific properties:

Theorem 2.1 *The following relationships hold:*

1. *A serial rough set model satisfies (D).*
2. *A reflexive rough set model satisfies (T).*
3. *A symmetric rough set model satisfies (B).*
4. *A transitive rough set model satisfies (4).*

5. A Euclidean rough set model satisfies (5).

Proof (1) In a serial rough set model, for any $x \in apr(A)$, we have $r(x) \subseteq A$ and $r(x) \neq \emptyset$. This implies $r(x) \cap A \neq \emptyset$., i.e., $x \in \overline{apr}(A)$. Thus, property (D) holds in a serial rough set model.

(2) In a reflexive rough set model, for any $x \in U$, xRx implies $x \in r(x)$. Suppose $x \in apr(A)$, which is equivalent to $r(x) \subseteq A$. Combining $x \in r(x)$ and $r(x) \subseteq A$, we have $x \in A$. Thus, property (T) holds.

(3) In a symmetric rough set model, suppose $x \in A$. By the symmetry of R, for all $y \in r(x)$, we have $x \in r(y)$, i.e., $x \in r(y) \cap A$. This implies that for all $y \in r(x)$, $y \in \overline{apr}(A)$. Hence, $r(x) \subseteq \overline{apr}(A)$. It means that $x \in apr(\overline{apr}(A))$. Therefore, property (B) holds.

(4) In a transitive rough set model, suppose $x \in apr(A)$, i.e., $r(x) \subseteq A$. Then, for all $y \in r(x)$, $r(y) \subseteq r(x) \subseteq A$, which is equivalent to say that for all $y \in r(x)$, $y \in apr(A)$. Thus, $r(x) \subseteq apr(A)$ and in turn $x \in apr(apr(A))$. That is, property (4) holds.

(5) Consider an Euclidean rough set model. Suppose $x \in \overline{apr}(A)$, i.e., $r(x) \cap A \neq \emptyset$. By the Euclidean property of R, for all $y \in r(x)$, $r(x) \subseteq r(y)$. Combining this with the assumption $r(x) \cap A \neq \emptyset$, we can conclude that for all $y \in r(x)$, $y \in \overline{apr}(A)$. This is equivalent to say $r(x) \subseteq \overline{apr}(A)$. Therefore, (5) holds.

The five properties of a binary relation, namely the serial, reflexive, symmetric, transitive and Euclidean properties, induce five properties of the approximation operators, namely

> serial: property (D) holds.
> reflexive: property (T) holds.
> symmetric: property (B) holds.
> transitive: property (4) holds.
> Euclidean: property (5) holds.

Combining these properties, it is possible to construct more rough set models. As a consequence of Theorem 2.1, we have:

Theorem 2.2 *The following hold:*

1. *A $T = KT$ rough set model satisfies (K),(T).*
2. *A $B = KB$ rough set model satisfies (K), (B).*
3. *A $S4 = KT4$ rough set model satisfies (K), (T), and (4).*
4. *A $S5 = KT5$ (Pawlak) rough set model satisfies (K), (T), and (5).*

We can construct a rough set model corresponding to a normal modal system. Note that S5 (Pawlak) model is the strongest model. Yao and Lin's work established the connection of (normal) modal logic and rough set theory.

Yao and Lin generalized their modal logic approaches for graded and probabilistic modal logic. Here we review their approach.

Yao and Lin's work has been expanded in various ways. They presented graded and probabilistic versions. Because the quantitative information about the degree of overlap of $r(x)$ and A is not considered in modal logic approach, such extensions are interesting.

Graded modal logics extend modal logic by introducing a family of graded modal operators $\Box_n$ and $\Diamond_n$, where $n \in N$ and N is the set of natural numbers; see Fattorosi-Barnaba et al. [9, 15, 16]. These operators can be interpreted as follows:

(gm6) $w \models \Box_n\phi$ iff $|r(w)| - |t(\phi) \cap r(w)| \le n$
(gm7) $w \models \Diamond_n\phi$ iff $|t(\phi) \cap r(w)| > n$,

where $|\cdot|$ denotes the cardinality of a set. $t(\phi)$ is the set of possible worlds in which ϕ is true and $r(w)$ is the set of possible worlds accessible from w.

The interpretations of $\Box_n\phi$ is that ϕ is false in at most n possible worlds accessible from w, and the interpretation of $\Diamond_n\phi$ is that ϕ is true in more than n possible worlds accessible from n.

Graded necessity and possibility operators are dual:

$$\Box_n\phi =_{\text{def}} \neg\Diamond_n\neg\phi$$
$$\Diamond_n\phi =_{\text{def}} \neg\Box_n\neg\phi.$$

If $n = 0$, they reduce to normal modal operators.

$$\Box\phi =_{\text{def}} \neg\Diamond\neg\phi$$
$$\Diamond\phi =_{\text{def}} \neg\Box\neg\phi.$$

Further, we introduce a new graded modal operator $\Diamond!_n$, defined as $\Diamond!_n\phi = \Box_0\neg\phi$ and $\Diamond!_n\phi = \Diamond_{n-1}\phi \wedge \neg\Diamond_n\phi$ for $n > 0$. The interpretation of $\Diamond!_n\phi$ is that ϕ is true in exactly n possible worlds accessible from w.

We can formalize graded modal systems. The basic graded modal logic is called Gr(K), in which axiom (K) is replaced by the following three axioms:

(GK1) $\Box_0(\phi \to \psi) \to (\Box_n\phi \to \Box_n\psi)$
(GK2) $\Box_n\phi \to \Box_{n+1}\phi$
(GK3) $\Box_0\neg(\phi \wedge \psi) \to ((\Diamond!_n\phi \wedge \Diamond!_m\psi) \to \Diamond_{n+m}(\phi \vee \psi))$

The graded modal logic Gr(T) is obtained by adding (GT) to Gr(K), and Gr(S5) is obtained by adding (G5) to Gr(T). The two axioms (GT) and (G5) are defined as follows:

(GT) $\Box_0\phi \to \phi$
(G5) $\Diamond_n\phi \to \Box_0\Diamond_n\phi.$

Based on graded modal logic, we introduce the notion of graded rough sets. Given the universe U and a binary relation R on U, a family of graded rough set operators are defined as:

$$\underline{apr}_n(A) = \{x \mid |r(x)| - |A \cap r(x)| \le n\}$$
$$\overline{apr}_n(A) = \{x \mid |A \cap r(x)| > n\}.$$

An element of U belongs to $\underline{apr}_n(A)$ if at most n of its R-related elements are not in A, and belongs to $\overline{apr}_n(A)$ if more than n of its R-related elements are in A. We establish a link between graded modal logics and rough sets:

(gs6) $t(\Box_n\phi) = \underline{apr}_n(t(\phi))$

(gs7) $t(\Diamond_n\phi) = \overline{apr}_n(t(\phi))$

We can interpret graded rough set operators in terms of graded modal operators. Independent of the types of binary relations, graded rough set operators satisfy the following properties:

(GL0) $\underline{apr}(A) = \underline{apr}_0(A)$

(GL1) $\underline{apr}_n(A) = \sim \overline{apr}_n(\sim A)$

(GL2) $\underline{apr}_n(A) = U$

(GL3) $\underline{apr}_n(A \cap B) \subseteq \underline{apr}_n(A) \cap \underline{apr}_n(A)$

(GL4) $\underline{apr}_n(A \cup B) \supseteq \underline{apr}_n(A) \cup \underline{apr}_n(A)$

(GL5) $A \subseteq B \Rightarrow \underline{apr}_n(A) \subseteq \underline{apr}_n(B)$

(GL6) $n \geq m \Rightarrow \underline{apr}_n(A) \supseteq \underline{apr}_m(A)$

(GU0) $\overline{apr}(A) = \overline{apr}_0(A)$

(GU1) $\overline{apr}_n(A) = \sim \underline{apr}_n(\sim A)$

(GU2) $\overline{apr}_n(\emptyset) = \emptyset$

(GU3) $\overline{apr}_n(A \cup B) \supseteq \overline{apr}_n(A) \cup \overline{apr}_n(B)$

(GU4) $\overline{apr}_n(A \cap B) \subseteq \overline{apr}_n(A) \cap \overline{apr}_n(B)$

(GU5) $A \subseteq B \Rightarrow \overline{apr}_n(A) \subseteq \overline{apr}_n(B)$

(GU6) $n \geq m \Rightarrow \overline{apr}_n(A) \subseteq \overline{apr}_m(A)$

Properties (GL0) and (GU0) show the relationship between graded rough set operators and normal rough set operators.

Properties (GL1)–(GL5) and (GU1)–(GU5) correspond to properties (AL1)–(AL5) and (AU1)–(AU5) of algebraic rough sets.

For properties (GL3) and (GU3), set equality is replaced by set inclusion.

Properties (GL6) and (GU6) characterize the relationships between graded modal operators. In fact, (GL6) corresponds to a graded version of (GK2) of graded modal logic. Properties corresponding to (GK1) and (GK3) can be easily constructed.

It is possible to construct different graded rough set models based on the properties satisfied by the binary relation. If the binary relation R is an equivalence relation, we obtain the graded version of Pawlak rough sets.

The operators in graded Pawlak rough sets satisfy properties corresponding to axioms of graded modal logic:

(GD) $\underline{apr}_0(A) \subseteq \overline{apr}_0(A)$

(GT) $\underline{apr}_0(A) \subseteq A$

(GB) $A \subseteq \underline{apr}_0(\overline{apr}_0(A))$

(G4) $\underline{apr}_n(A) \subseteq \underline{apr}_0(\underline{apr}_n(A))$

(G5) $\overline{apr}_n(A) \subseteq \underline{apr}_0(\overline{apr}_n(A))$

We turn to probabilistic rough sets. Although we only use the *absolute* number of possible worlds accessible from a world w and in which a proposition ϕ is true (false) in the definition of graded modal operators, the size of $r(w)$ is not considered. In probabilistic modal logic, all such information will be used.

Let (W, R) be a frame. For each $w \in W$, we define a probabilistic function $P_w : \Phi \to [0, 1]$:

$$P_w(\phi) = \frac{|t(\phi) \cap r(w)|}{|r(w)|}$$

where $t(\phi)$ is the set of possible worlds in which ϕ is true, and $r(w)$ is the set of possible worlds accessible from w. Here, we implicitly assume that R is at least serial, i.e., for all $w \in W$, $|r(w)| \geq 1$. Then, we define a family of probabilistic modal logic operators for $\alpha \in [0, 1]$:

$$\begin{aligned}
\text{(pm6)} \quad & w \models \Box_\alpha \phi \text{ iff } P_w(\phi) \geq 1 - \alpha \\
& \text{iff } \frac{|t(w) \cap r(w)|}{|r(w)|} \geq 1 - \alpha \\
\text{(pm7)} \quad & w \models \Diamond_\alpha \phi \text{ iff } P_w(\phi) > \alpha \\
& \text{iff } \frac{|t(w) \cap r(w)|}{|r(w)|} > \alpha
\end{aligned}$$

These probabilistic modal operators are dual, i.e.,

$$\Box_\alpha \phi =_{\text{def}} \neg \Diamond_\alpha \neg \phi$$
$$\Diamond_\alpha \phi =_{\text{def}} \neg \Box_\alpha \neg \phi.$$

If $\alpha = 0$, then they agree to normal modal operators:

$$\Box_0 \phi =_{\text{def}} \Box \phi$$
$$\Diamond_0 \phi =_{\text{def}} \Diamond \phi.$$

The definition of probabilistic modal operators is consistent with that of Murai et al. [31, 33]. It is also a special case of the probabilistic Kripke model of Fattorosi-Barnaba and Amati in [14].

In fact, the probabilistic modal operators are related to the graded modal operators. If both sides of inequalities in (gm6) and (gm7) are divided by $|r(w)|$, and $n/|r(w)|$ is replaced by α, the probabilistic modal operators are obtained.

But, these operators are different. Consider two possible worlds $w, w' \in W$ with $|r(w) \cap t(\phi)| = |r(w') \cap t(\phi)| = 1$ and $|r(w)| \neq |r(w')|$. We have:

$$w \models \Diamond_0 \phi$$
$$w' \models \Diamond_0 \phi$$

and

$$w \models \neg \Diamond_n \phi$$
$$w' \models \neg \Diamond_n \phi$$

for $n \geq 1$. That is, evaluations of $\Diamond_n \phi$ are the same in both worlds w and w'. The difference in the size of $r(w)$ and $r(w')$ is reflected by operators $\Box_n$. On the other hand, since $1/|r(x)| \neq 1/|r(y)|$, evaluations of both $\Diamond_\alpha \phi$ and $\Box_\alpha$ will be different in worlds w and w'.

We can then define probabilistic rough sets. Let U be the universe and R be a binary relation on U. We define a family of probabilistic rough set operators.

$$\underline{apr}_\alpha(A) = \{x \mid \frac{|A \cap r(x)|}{|r(x)|} \geq 1 - \alpha\}$$

$$\overline{apr}_\alpha(A) = \{x \mid \frac{|A \cap r(x)|}{|r(x)|} > \alpha\}.$$

With this definition, we can establish the connections between probabilistic modal logic and probabilistic rough sets:

(ps6) $t(\square_\alpha \phi) = \underline{apr}_\alpha(t(\phi))$
(ps7) $t(\lozenge_\alpha \phi) = \overline{apr}_\alpha(t(\phi))$.

By definition, for a serial binary relation and $\alpha \in [0, 1]$, probabilistic rough set operators satisfy the following properties:

(PL0) $apr(A) = \underline{apr}_0(A)$
(PL1) $\underline{apr}_\alpha(A) = \sim \overline{apr}_\alpha(\sim A)$
(PL2) $\underline{apr}_\alpha(U) = U$
(PL3) $\underline{apr}_\alpha(A \cap B) \subseteq \underline{apr}_\alpha \cap \underline{apr}_\alpha(B)$
(PL4) $\underline{apr}_\alpha(A \cup B) \subseteq \underline{apr}_\alpha \cup \underline{apr}_\alpha(B)$
(PL5) $A \subseteq B \Rightarrow \underline{apr}_\alpha(A) \subseteq \underline{apr}_\alpha(B)$
(PL6) $\alpha \geq \beta \Rightarrow \underline{apr}_\alpha(A) \supseteq \underline{apr}_\beta(A)$
(PU0) $\overline{apr}(A) = \overline{apr}_0(A)$
(PU1) $\overline{apr}_\alpha(A) = \sim \underline{apr}_\alpha(\sim A)$
(PU2) $\overline{apr}_\alpha(\emptyset) = \emptyset$
(PU3) $\overline{apr}_\alpha(A \cup B) \supseteq \overline{apr}_\alpha(A) \cup \overline{apr}_\alpha(B)$
(PU4) $\overline{apr}_\alpha(A \cap B) \subseteq \overline{apr}_\alpha(A) \cap \overline{apr}_\alpha(B)$
(PU5) $A \subseteq B \Rightarrow \overline{apr}_\alpha(A) \subseteq \overline{apr}_\alpha(B)$
(PU6) $\alpha \geq \beta \Rightarrow \overline{apr}_\alpha(A) \subseteq \overline{apr}_\beta(A)$

They are counterparts of the properties of graded rough set operators. Moreover, for $0 \leq \alpha < 0.5$,

(PD) $\underline{apr}_\alpha(A) \subseteq \overline{apr}_\alpha(A)$

which may be interpreted as a probabilistic version of axiom (D).

Probabilistic rough sets were first introduced by Wong and Ziarko [59] using probabilistic functions. Murai et al. also proposed similar probabilistic modal operators in [30, 32].

Scott-Montague models—or minimal models [8]—are a generalization of Kripke models. A Scott-Montague model is a triple

$$M = \langle W, N, v \rangle$$

where W is a non-empty set of possible worlds and v is a truth valuation function used in Kripke models. Instead of accessibility relation R, the Scott-Montague model uses function N from W to 2^{2^W}. Truth conditions of modal sentences in the Scott-Montague model M are given by

$M, w \models \Box p \Leftrightarrow \| p \|^M \in N(w)$.
$M, w \models \Diamond p \Leftrightarrow (\| p \|^M)^c \notin N(w)$.

Conditions of N are considered such that

(m) $X \cap Y \in N(w) \Rightarrow X \in N(w)$ and $Y \in N(w)$,
(c) $X, Y \in N(w) \Rightarrow X \cap Y \in N(w)$,
(n) $W \in N(w)$,
(d) $X \in N(w) \Rightarrow X^c \notin N(w)$,
(t) $X \in N(w) \Rightarrow w \in X$,
(4) $X \in N(w) \Rightarrow \{x \in W \mid X \in N(x)\} \in N(w)$,
(5) $X \notin N(w) \Rightarrow \{x \in W \mid X \notin N(x)\} \in N(w)$.

For every Kripke model $M_K = \langle W, R, v \rangle$ there is equivalent Scott-Montague model $M_{SM} = \langle W, N, v \rangle$ such that $M_K, w \models p$ iff $M_{SM}, w \models p$ for any sentence $p \in \mathcal{L}_{ML}(\mathcal{P})$ and any possible world $w \in W$. From accessibility relation R in Kripke model M, function N in Scott-Montague model M_{SM} is constructed by

$x \in N(w) \Leftrightarrow \{x \in W \mid wRx\}$

for every w. Function N in M_{SM} satisfies above conditions (m), (c), and (n), and the following condition:

(a) $X \in N(w) \Leftrightarrow \bigcup N(w) \subseteq X$,

where $\bigcup N(w)$ is the intersection of all sets in $N(w)$. Kripke models are thus special cases of Scott-Montague models.

Smallest classical modal logic E is proved to be both sound and complete with respect to the class of all Scott-Montague models, where E contains schema $D\Diamond$: $\Diamond p \leftrightarrow \neg\Box\neg p$ and rule of inference

RE: from $p \leftrightarrow q$ infer $\Box p \leftrightarrow \Box q$

with rules and axiom schemata of propositional logic.

Each condition of N corresponds to axiom schema such that

(M): $\Box(p \wedge q) \to (\Box p \wedge \Box q)$,
(C): $(\Box p \wedge \Box q) \to \Box(p \wedge q)$,
(N): $\Box\top$,
(D): $\Box p \to \Diamond p$,
(T): $\Box p \to p$,
(4): $\Box p \to \Box\Box p$,
(5): $\Diamond p \to \Box\Diamond p$.

Scott-Montague semantics can generalize the modal logic approach to rough sets based on Kripke semantics. In Chap. 11, we will discuss a modal characterization of visibility and focus in granular reasoning by means of Scott-Montague semantics.

2.4   Rough Set Logics

It is necessary to advance a logical system based on rough sets for practical applications. Such a logic is called a *rough set logic*. The first approach to a rough set logic was established by Düntsch [11] in 1997. Here, we quickly review his system and related ones.

Düntsch developed a propositional logic for rough sets inspired by the topological construction of rough sets using Boolean algebras. His work is based on the fact that the collection of all subsets of a set forms a Boolean algebra under the set-theoretic operation, and that the collection of rough sets of an approximation space is a regular double Stone algebra. Thus, we can assume that regular double Stone algebras can serve as a semantics for a logic for rough sets.

To understand his logic, we need some concepts. A *double Stone algebra DSA* is denoted by $\langle L, +, \cdot, *.+, 0, 1 \rangle$ with the type $\langle 2, 2, 1, 1, 0, 0 \rangle$ satisfying the following conditions:

(1) $\langle L, +, \cdot, 0, 1 \rangle$ is a bounded distributed lattice.
(2) x^* is the pseudocomplement of x, i.e., $y \leq x^* \Leftrightarrow y \cdot x = 0$.
(3) x^+ is the dual pseudocomplement of x, i.e., $y \geq x^+ \Leftrightarrow y + x = 1$.
(4) $x^* + x^{**} = 1, x^+ \cdot x^{++} = 0$.

DSA is called *regular* if it satisfies the additional condition: $x \cdot x^+ \leq x + x^*$. Let B be a Boolean algebra, F be a filter on B, and $\langle B, F \rangle = \{\langle a, b \rangle \mid a, b \in B, a \leq b, a + (-b) \in F\}$.

We define the following operations on $\langle B, F \rangle$ as follows:

$$\langle a, b \rangle + \langle c, d \rangle = \langle a + c, b + d \rangle,$$
$$\langle a, b \rangle \cdot \langle c, d \rangle = \langle a \cdot c, b \cdot d \rangle,$$
$$\langle a, b \rangle^* = \langle -b, -b \rangle,$$
$$\langle a, b \rangle^+ = \langle -a, -a \rangle.$$

If $\langle U, R \rangle$ is an approximation space, the classes of R can be viewed as a complete subalgebra of the Boolean algebra $B(U)$. Conversely, any atomic complete subalgebra B of $B(U)$ yields an equivalence relation R on U by the relation: $xRy \Leftrightarrow x$ and y are contained in the same atom of B, and this correspondence is bijective.

If $\{a\} \in B$, then for every $X \subseteq U$ we have: If $a \in \underline{R}X$, then $a \in X$, and the rough sets of the corresponding approximation space are the elements of the regular double Stone algebra $\langle B, F \rangle$, where F is the filter of B which is generated by the union of the singleton elements of B.

Based on the construction of regular double Stone algebras, Düntsch proposed a propositional rough set logic *RSL*. The language $\mathcal{L}$ of *RSL* has two binary connectives $\wedge$ (conjunction), $\vee$ (disjunction), two unary connectives $*, ^+$ for two types of negation, and the logical constant $\top$ for truth.

Let P be a non-empty set of propositional variables. Then, the set **Fml** of formulas with the logical operators constitutes an absolutely free algebra with a type $\langle 2, 2, 1, 1, 0 \rangle$. Let W be a set and $B(W)$ be a Boolean algebra based on W.

Then, a *model M* of L is seen as a pair (W, v), where $v : P \to B(W) \times B(W)$ is the *valuation function* for all $p \in P$ satisfying: if $v(p) = \langle A, B \rangle$, then $A \subseteq B$. Here,

$v(p) = \langle A, B \rangle$ states that p holds at all states of A and does not hold at any state outside B.

Düntsch relates the valuation to Lukasiewicz's three-valued logic by the following construction. For each $p \in P$, let $v_p : W \to \mathbf{3} = \{0, \frac{1}{2}, 1\}$. $v : P \to B(W) \times B(W)$ is defined as follows: $v(p) = \langle \{w \in W : v_p(w) = 1\}, \{w \in W : v_p(w) \neq 0\} \rangle$. In addition, Düntsch connected the valuation and rough sets as follows:

$v_p(w) = 1$ if $w \in A$,
$v_p(w) = \frac{1}{2}$ if $w \in B \setminus A$,
$v_p(w) = 0$ otherwise.

Given a model $M = (W, v)$, the *meaning function* $\mathrm{mng} : \mathbf{Fml} \to B(W) \times B(W)$ is defined to give a valuation of arbitrary formulas in the following way:

$\mathrm{mng}(\top) = \langle W, W \rangle$,
$\mathrm{mnf}(\to p) = \langle W, W \rangle$,
$\mathrm{mng}(p) = v(p)$ for $p \in P$.

If $\mathrm{mng}(\phi) = \langle A, B \rangle$ and $\mathrm{mng}(\psi) = \langle C, D \rangle$, then

$\mathrm{mng}(\phi \wedge \psi) = \langle A \cap C, B \cap D \rangle$,
$\mathrm{mng}(\phi \vee \psi) = \langle A \cup C, B \cup D \rangle$,
$\mathrm{mng}(\phi^*) = \langle -B, -B \rangle$,
$\mathrm{mng}(\phi^+) = \langle -A, -A \rangle$.

Here, $-A$ denotes the complement of A in $B(W)$. We can understand that the meaning function assigns the meaning to formulas.

The class of all models of L is denoted by **Mod**. A formula A *holds* in a model $M = \langle W, v \rangle$, written $M \models A$, if $\mathrm{mng}(A) = \langle W, W \rangle$. A set Γ of sentences *entails* a formula A, written $\Gamma \vdash A$, if every model of Γ is a model of A.

We can define additional operations on **Fml** by

$A \to B = A^* \vee B \vee (A^+ \wedge B^{**})$
$A \leftrightarrow B = (A \to B) \wedge (B \to A)$

Düntsch proved several technical results including completeness.

Theorem 2.3 *If $M = \langle W, c \rangle \in$ **Mod** and $\phi, \psi \in$ **Fml**, then*

1. $M \models \phi \leftrightarrow \psi$ iff $\mathrm{mng}(\phi) = \mathrm{mng}(\psi)$
2. $M \models\to p \leftrightarrow \phi$ iff $M \models \phi$.

Theorem 2.4 states completeness, compactness, and Beth definability property of *RSL*.

Theorem 2.4 *The following hold:*

1. *RSL has a finitely complete and strongly sound Hilbert style axiom system.*
2. *RSL has a compactness theorem.*
3. *RSL does not have the Beth definability property.*

The implication in *RSL* is interesting, since it has no intuitive appeal. It is thus promising to extend *RSL* with another implication. Such extensions can be found in Akama et al. in [2, 3], which relate rough set logics to Heyting-Brouwer logic and its sub-logic.

2.5 Logics for Reasoning About Knowledge

A first approach to connect rough set theory and modal logic was due to Orlowska's
series of papers [39–41]. She implicitly showed the relation of Pawlak's rough sets
and modal logic S5. However, she did more than this. Here, we review her logics for
reasoning about knowledge.

Orlowska proposed a logic with knowledge operators which are relative to indiscernibility relations associated with agents, with a semantics based on rough sets.
Orlowska's approach is the following three intuitions:

1. Knowledge of an agent about a predicate F can be reflected by the ability of the
   agent to classify objects as instances or non-instances of F.
2. Knowledge of an agent about a sentence F can be reflected by the ability of the
   agent to classify states into those in which F is true and those in which F is false.
3. With each agent there is associated to an indiscernibility relation, and the agent
   decides membership of objects or states up to this indiscernibility. As a consequence, knowledge operators are relative to indiscernibility.

Let U be a universe (of states or objects) and let AGT be a set of agents. For each
$a \in AGT$, let $ind(a) \subseteq U \times U$ be an indiscernibility relation corresponding to agent
a.

For a set A of agents, we define indiscernibility $ind(A)$ as follows:

$(s, t) \in ind(A)$ iff $(s, t) \in ind(a)$ for all $a \in A$
$ind(\emptyset) = U \times U$.

Indiscernibility relations are equivalence relation (reflexive, symmetric and transitive) or similarity relations (reflexive and symmetric). Below, we confine ourselves
to equivalence relations.

We here state the following properties of indiscernibility relations.

Proposition 2.3 *Indiscernibility relations satisfy the following:*

1. $ind(A \cup B) = ind(A) \cap ind(B)$
2. $(ind(A) \cup ind(B))^* \subseteq ind(A \cap B)$
3. $A \subseteq B$ implies $ind(B) \subseteq ind(A)$

Here, $(ind(A) \cup ind(B))^* = ind(\{A, B\}^*)$ *is the set of all finite sequences with
elements from the set* $\{A, B\}$.

Proposition 2.4 *The family* $\{ind(A)\}_{A \subseteq AGT}$ *is a lower semilattice in which*
$ind(AGT)$ *is the zero element.*

Now, we assume a subset X of the universe U and an indiscernibility relation
$ind(A)$ for a certain set of agents. Since agents recognize elements of U up to $ind(A)$,
they grasp X within the limit of tolerance determined by lower and upper approximation of X.

The lower approximation $\underline{ind}(A)X$ of X with respect to $ind(A)$ is the union of those equivalence classes determined by $ind(A)$ which are included in X. The upper approximation $\overline{ind}(A)X$ of X with respect to $ind(A)$ is the union of those equivalence classes determined by $ind(A)$ which have an element in common with X.

For non-transitive indiscernibility relations the respective definitions of approximations are obtained by taking similarity classes instead of equivalence classes.

Proposition 2.5 $\underline{ind}(A)X$ and $\overline{ind}(A)X$ satisfy the following:

1. $x \in \underline{ind}(A)X$ iff for all $t \in U$ if $(x, t) \in ind(A)$, then $t \in X$.
2. $x \in \overline{ind}(A)X$ iff there is $t \in U$ such that $(x, t) \in ind(A)$ and $t \in X$.

Proposition 2.6 The following relations hold.

1. $\underline{ind}(A)\emptyset = \emptyset$, $\overline{ind}(A)U = U$
2. $\underline{ind}(\emptyset)X = \emptyset$ for $X \neq U$, $\underline{ind}(\emptyset)U = U$,
   $\overline{ind}(\emptyset)X = U$ for $X \neq \emptyset$, $\overline{ind}(\emptyset)\emptyset = \emptyset$,
3. $\underline{ind}(A)X \subseteq X \subseteq \overline{ind}(A)X$
4. $\underline{ind}(A)\underline{ind}(A)X = \underline{ind}(A)X$, $\underline{ind}(A)\overline{ind}(A)X = \overline{ind}(A)X$,
   $\overline{ind}(A)\underline{ind}(A)X = \underline{ind}(A)X$, $\overline{ind}(A)\underline{ind}(A)X = \underline{ind}(A)X$
5. $X \subseteq Y$ implies $\underline{ind}(A)X \subseteq \underline{ind}(A)Y$ and $\overline{ind}(A)X \subseteq \overline{ind}(A)Y$
6. $ind(A) \subseteq ind(B)$ implies $\underline{ind}(B)X \subseteq \underline{ind}(A)X$ and $\overline{ind}(A)X \subseteq \overline{ind}(B)X$ for any $X \subseteq U$.

Proposition 2.7 The following relations for complement hold.

1. $\overline{ind}(A)X \cup \underline{ind}(A)(-X) = U$
2. $\overline{ind}(A)X \cap \underline{ind}(A)(-X) = \emptyset$
3. $-\overline{ind}(A)X = \underline{ind}(A)(-X)$
4. $-\underline{ind}(A)X = \overline{ind}(A)(-X)$.

Proposition 2.8 The following relations for union and intersection hold.

1. $\underline{ind}(A)X \cup \underline{ind}(A)Y \subseteq \underline{ind}(A)(X \cup Y)$
2. $X \cap Y = \emptyset$ implies $\underline{ind}(A)(X \cup Y) = \underline{ind}(A)X \cup \underline{ind}(A)Y$
3. $\underline{ind}(A)(X \cap Y) = \underline{ind}(A)X \cap \underline{ind}(A)Y$
4. $\overline{ind}(A)(X \cup Y) = \overline{ind}(A)X \cap \overline{ind}(A)Y$
5. $\overline{ind}(A)(X \cap Y) \subseteq \overline{ind}(A)X \cap \overline{ind}(A)Y$.

A set $X \subseteq U$ is said to be

A-definable	iff $\underline{ind}(A)X = X = \overline{ind}(A)X$ or $X = \emptyset$
roughly A-definable	iff $\underline{ind}(A)X \neq \emptyset$ and $\overline{ind}(A)X \neq U$
internally A-definable	iff $\underline{ind}(A)X = \emptyset$
externally A-definable	iff $\overline{ind}(A)X = U$
totally A-non-definable	iff it is internally A-non-definable and externally A-non-definable

We can define sets of A-positive, A-negative and A-borderline instances of a set $X \subseteq U$.

$$POS(A)X = \underline{ind}(A)X,$$
$$NEG(A)X = \overline{-ind}(A)X,$$
$$BOR(A)X = \overline{ind}(A)X - \underline{ind}(A)X$$

Intuitively, if $s \in POS(A)X$, then in view of agents from A element s is a member of X. If $s \in NEG(A)X$, then in view of agents from A element s is not a member of X. $BOR(A)X$ is the range of uncertainty. Element $s \in BOR(A)X$ whenever agents from A cannot decide whether s is a member of X or not.

For these sets, we have the following propositions:

Proposition 2.9 *We have the following:*

1. $POS(A)X$, $NEG(A)X$, $BOR(A)X$ *are pairwise disjoint.*
2. $POS(A)X \cup NEG(A)X \cup BOR(A)X = U$.
3. $POS(A)X$, $NEG(A)X$, $BOR(A)X$ *are A-definable.*

Proposition 2.10 *We have the following:*

1. $A \subseteq B$ *implies* $POS(A)X \subseteq POS(B)X$, $NEG(A)X \subseteq NEG(B)X$, $BOR(B)$ $X \subseteq BOR(A)X$.
2. $ind(A) \subseteq ind(B)$ *implies* $POS(B)X \subseteq POS(A)X$, $NEG(B)X \subseteq NEG(A)$ X, $BOR(A)X \subseteq BOR(B)$.

Proposition 2.11 *We have the following:*

1. $POS(A)X \subseteq X$, $NEG(A)X \subseteq -X$
2. $POS(A)\emptyset = \emptyset$, $NEG(A)U = \emptyset$
3. $POS(\emptyset)X = \emptyset$ *if* $X \neq U$, $POS(\emptyset)U = U$
4. $NEG(\emptyset)X = \emptyset$ *if* $X \neq \emptyset$, $NEG(\emptyset)\emptyset = U$
5. $X \subseteq Y$ *implies* $POS(A)X \subseteq POS(A)Y$, $NEG(A)Y \subseteq NEG(A)X$.

Proposition 2.12 *We have the following:*

1. $POS(A)X \cup POS(A)Y \subseteq POS(A)(X \cup Y)$
2. *If* $X \cap Y = \emptyset$ *then* $POS(A)(X \cup Y) = POS(A)X \cup POS(B)Y$
3. $POS(A)(X \cap Y) = POS(A)X \cap POS(A)Y$
4. $NEG(A)(X \cup Y) = NEG(A)X \cap NEG(A)Y$
5. $NEG(A)X \cup NEG(A)Y \subseteq NEG(A)(X \cup Y)$
6. $NEG(A)(-X) = POS(A)X$.

Proposition 2.13 *We have the following:*

1. $BOR(A)(X \cup Y) \subseteq BOR(A)X \cup BOR(A)Y$
2. $X \cap Y = \emptyset$ *implies* $BOR(A)(X \cup Y) = BOR(A)X \cup BOR(A)Y$
3. $BOR(A)(X \cap Y) \subseteq BOR(A)X \cap BOR(A)Y$
4. $BOR(A)(-X) = BOR(A)X$.

Proposition 2.14 *We have the following:*

1. $POS(A)X \cup POS(B)X \subseteq POS(A \cup B)X$
2. $POS(A \cap B)X \subseteq POS(A)X \cap POS(B)X$
3. $NEG(A)X \cup NEG(B)X \subseteq NEG(A \cup B)X$
4. $NEG(A \cap B)X \subseteq NEG(A)X \cap NEG(B)X.$

Proposition 2.15 *We have the following:*

1. $POS(A)POS(A)X = POS(A)X$
2. $POS(A)NEG(A)X = NEG(A)X$
3. $NEG(A)NEG(A)X = -NEG(A)X$
4. $NEG(A)POS(A)X = -POS(A)X.$

We define a family of knowledge operators $K(A)$ for $A \in AGT$:

$$K(A)X = POS(A)X \cup NEG(A)X.$$

Intuitively, $s \in K(A)X$ whenever s can be decided by agents from A to be A-positive or A-negative instance of X.

Proposition 2.16 *We have the following:*

1. $K(A)\emptyset, K(A)U = U$
2. $K(\emptyset)X = \emptyset$ if $X \neq U$
3. $ind(A) \subseteq ind(B)$ implies $K(B)X \subseteq K(A)X$ for all $X \subseteq U$
4. $A \subseteq B$ implies $K(A)X \subseteq K(B)X$ for all $X \subseteq U$
5. If X is A-definable, then $K(A)X = U$.

We say that knowledge of agents A about X is:

$$
\begin{array}{ll}
\text{complete} & \text{iff } K(A)X = U \\
\text{incomplete} & \text{iff } BOR(A)X \neq \emptyset \\
\text{rough} & \text{iff } POS(A)X, NEG(A)X, BOR(A)X \neq \emptyset \\
\text{pos-empty} & \text{iff } POS(A)X = \emptyset \\
\text{neg-empty} & \text{iff } NEG(A)X = \emptyset \\
\text{empty} & \text{iff it is pos-empty and neg-empty.}
\end{array}
$$

If knowledge of A about X is complete, then A can discern X from its complement. Every A has a complete knowledge about any A-definable set, in particular, about $\emptyset$ and U. The fact that knowledge of any agent about the whole universe is complete should not be considered to be a paradox.

A predicate whose extension equals U provides a trivial, in a sense, information. In any particular example U represents the set of "all things perceivable by agents". However, if U consists of all formulas of the predicate calculus, and $X \subseteq U$ is the set of all the valid formulas, then clearly not every agent has the complete knowledge about X, although he has the complete knowledge about U.

Observe that $X \subseteq Y$ does not imply $K(A)X \subseteq K(A)Y$, and $K(A)X$ is not necessarily included in X. By these facts, we can avoid the well-known paradoxes of epistemic logic, i.e., *the problem of logical omniscience*, where all the formulas known by anyone are valid, and every agent knows all the logical consequences of his knowledge.

Proposition 2.17 *The following conditions are equivalent:*

1. $K(A)X$ *is complete.*
2. X *is A-definable.*
3. $BOR(A)X = \emptyset$.
4. $POS(A)X = -NEG(A)X$.

It follows that if agents A have complete knowledge about X, then they can tell X from its complement.

Proposition 2.18 *The following conditions are equivalent:*

1. $K(A)X$ *is rough.*
2. X *is roughly A-definable.*
3. $\emptyset \neq BOR(A)X \neq U$.
4. $POS(A)X \subseteq -NEG(A)X$.

Proposition 2.19 *The following conditions are equivalent:*

1. $K(A)X$ *is pos-empty.*
2. X *is internally A-non-definable.*
3. $K(A)X = POS(A)X$.
4. $BOR(A)X = -NEG(A)X$.

Proposition 2.20 *The following conditions are equivalent:*

1. $K(A)X$ *is neg-empty.*
2. X *is externally A-non-definable.*
3. $K(A)X = POS(A)X$.
4. $BOR(A)X = -POS(A)X$.

Proposition 2.21 *The following conditions are equivalent:*

1. $K(A)X$ *is empty.*
2. X *is totally A-non-definable.*
3. $BOR(A)X = U$.

Proposition 2.22 *The following hold:*

1. $K(A)X \subseteq K(B)X$ *for all* $X \subseteq U$ *implies* $ind(B) \subseteq ind(A)$.
2. $ind(A) = ind(B)$ *iff* $K(A)X = K(B)X$ *for all* $X \subseteq U$.
3. $ind(A) \subseteq ind(B)$ *implies* $K(B)X \subseteq POS(B)K(A)X$ *and* $POS(A)K(B)X \subseteq K(A)X$.
4. $ind(A)$ *is the identity on* U *iff* $K(A)X$ *is complete for all* X.

Proposition 2.23 *The following hold:*

1. $K(A)X = K(A)(-X)$
2. $K(A)K(A)X = U$
3. $K(A)X \cup K(B)X \subseteq K(A \cup B)X$
4. $K(A \cap B)X \subseteq K(A)X \cap K(B)X.$

Next, we discuss the independence of agents. A set A of agents is said to be *dependent* iff there is $B \subset A$ such that $K(A)X = K(B)X$ for all $X \subseteq U$. A set A is *independent* if it is not dependent.

Proposition 2.24 *The following conditions are equivalent:*

1. *A is independent.*
2. *For every $B \subset A$ there is $X \subseteq U$ such that $K(B)X \subset K(A)X$.*

Proposition 2.25 *The following hold:*

1. *If A is independent, then every of its subsets is independent.*
2. *If A is dependent, then every of its supersets is dependent.*

Proposition 2.26 *If AGT is independent, then for any $A, B \subseteq AGT$, the following conditions are satisfied:*

1. $K(A)X \subseteq K(B)X$ *for all X implies $A \subseteq B$.*
2. $ind(A \cap B) = (ind(A) \cup ind(B))^*.$

The intuitive meaning of independence of a set of agents is that if we drop some agents from the independent set, then knowledge of the group of the remaining agents is less than knowledge of the whole group.

Similarly, if a set is dependent, then some of its elements are superfluous, we can drop them without changing knowledge of the group.

We say that a set B of agents is *superfluous* in a set A iff for all $X \subseteq U$, we have $K(A - B)X = K(A)X.$

Proposition 2.27 *The following hold:*

1. *If a set A is dependent, then there is $B \subset A$ such that B is superfluous in AGT.*
2. *A set S is dependent iff there is $B \subset A$ such that B is superfluous in A.*

Next, we concern joint knowledge and common knowledge. Knowledge relative to indiscernibility $ind(A \cup B)$ can be considered to be a joint knowledge of A and B. $ind(A \cup B)$ is not greater than indiscernibility relations $ind(A)$ and $ind(B)$.

Proposition 2.28 $K(A)X, K(B)X \subseteq K(A \cup B)X.$

Hence, a joint knowledge of a group of agents is not less than knowledge of any member of the group.

Knowledge relative to indiscernibility $ind(A \cap B)$ can be considered to be a common knowledge of A and B. To discuss common knowledge, we have to admit non-transitive indiscernibility relations. Here, we introduce the following notation:

$ind(A) \circ ind(B) = ind(AB)$, where $\circ$ is the composition of relations,
$(ind(A) \cup ind(B))^* = ind(\{A, B\}^*)$.

Here, $\{A, B\}^*$ is the set of all finite sequences with elements from the set $\{A, B\}$.

For $S \in \{A, B\}^*$, $ind(S)$ is the composition of $ind(A)$ for all the elements A of sequence S.

Proposition 2.29 *The following hold:*

1. $ind(A), ind(B) \subseteq ind(AB)$
2. $ind(\{A, B\}^*) \subseteq ind(A \cap B)$
3. If $ind(AB) = ind(BA)$, then $ind(\{A, B\}^*) = ind(AB)$
4. If $ind(A) \subseteq ind(B)$, then $ind(AC) \subseteq ind(BC)$ and $ind(CA) \subseteq ind(CB)$ for any $C \subseteq AGT$
5. If AGT is independent, then $ind(\{A, B\}^*) = ind(A \cap B)$.

Observe that for $S \in \{A, B\}^*$, the relation $ind(S)$ is not necessarily transitive.

Proposition 2.30 $K(A \cap B)X \subseteq K(S)X$ *for any* $S \in \{A, B\}^*$.

Hence, common knowledge of A and B is included in the knowledge relative to composition of relations $ind(A)$ and $ind(B)$.

Orlowska defined a propositional language with a family of relative knowledge operators. Each operator is determined by a set of parameters interpreted as knowledge agents.

Let $CONAGT$ be a set of constants which are to be interpreted as sets of agents. We define the set $EXPAGT$ of agent expressions:

$CONAGT \subseteq EXPAGT$,
$A, B \in EXPAGT$ implies $-A, A \cup B, A \cap B \in EXPAGT$.

Let $VARPROP$ be a set of propositional variables. The set FOR of formulas is the smallest set satisfying the following conditions:

$VARPROP \subseteq FOR$
$F, G \in FOR$ implies $\neg F, F \vee G, F \wedge G, F \rightarrow G, F \leftrightarrow G \in FOR$
$A \in EXPAGT, F \in FOR$ imply $K(A)F \in FOR$

The set FOR is closed with respect to classical propositional connectives and knowledge operators determined by agent expressions.

Let an *epistemic system* $E = (U, AGT, \{ind(A)\}_{A \in AGT})$ be given. By a *model*, we mean a system $M = (E, m)$, where m is a meaning function assigning sets of states to propositional variables, sets of agents to agent constants, and moreover m satisfies the following conditions:

$m(p) \subseteq U$ for $p \in VARPROP$,
$m(A) \subseteq AGT$ for $A \in CONAGT$,
$m(A \cup B) = m(A) \cup m(B)$,
$m(A \cap B) = m(A) \cap m(B), m(-A) = -m(A)$.

We define inductively a family of set $ext_M\ F$ (extension of formula F in model M) for any $F \in FOR$:

$ext_M\ p = m(p)$ for $p \in VARPROP$,
$ext_M\ (\neg F) = -ext_M\ F$,
$ext_M\ (F \vee G) = ext_M\ F \cup ext_M\ G$,
$ext_M\ (F \wedge G) = ext_M\ F \cap ext_M\ G$,
$ext_M\ (F \rightarrow G) = ext_M\ (\neg \vee G)$,
$ext_M\ (F \leftrightarrow G) = ext_M\ ((F \rightarrow G) \wedge (G \rightarrow F))$,
$ext_M\ K(A)F = K(m(A))ext_M\ F$.

We say that a formula F is *true in a model* M ($\models_M F$) iff $ext_M\ F = U$ and a formulas F is *valid* ($\models F$) iff it is true in all models.

Observe that formulas of the form $K(A)F \rightarrow F$ and $(F \rightarrow G) \wedge K(A)F \rightarrow K(A)G$ are not valid. This means that if F is known by an agent, then F is not necessarily true, and agents do not know all the consequences of their knowledge. Thus, the system can avoid well-known paradoxes in epistemic logic.

In the following, we list some facts about knowledge of agents which can be expressed in the logic.

Proposition 2.31 *The following hold:*

1. $\models (F \leftrightarrow G)$ *implies* $\models (K(A)F \leftrightarrow K(A)G)$
2. $\models F$ *implies* $\models K(A)F$
3. $\models K(A)F \rightarrow K(A)K(A \cup N)F$
4. $\models K(A \cup B)K(A)F \rightarrow K(A \cup B)F$
5. $\models (K(A)F \vee K(B)F) \rightarrow K(A \cup B)F$
6. $\models K(A \cap B)F \rightarrow K(A)F \wedge K(B)F$
7. $\models K(A)F \leftrightarrow K(A)(\neg F)$
8. $\models K(A)(K(A)F \rightarrow F)$.

Here, (4) says that knowledge of a group of agents exceeds knowledge of a part of the group. (7) results from the fact that agents A can tell extension of F from its complement iff they can tell extension of $\neg F$ from its complement.

Observe that $\models F$ implies $\models K(A)F$. This fact is often considered to be a paradox of ideal knowers. However, it seems to be less paradoxical by using the interpretations of knowledge as the ability to decide the membership question.

It follows from the fact that the whole universe U is A-definable for any A. In other words, whatever a perception ability of A is (whatever $ind(A)$ is), equivalence classes of all the elements from U over U.

Orlowska's logic is significantly different from standard *epistemic logic* (cf. Hintikka [20], Fagin et al. [12], Halpern et al. [18, 19]), although it is based on Kripke semantics. Her logic can in fact overcome several defects of epistemic logic, and can be considered to be an interesting alternative for a logic of knowledge.

Orlowska only developed the semantics for her logics, and complete axiomatizations are open. However, for applications to real problems, we have to investigate a proof theory for logics for reasoning about knowledge.

In Orlowska's approach, approximation operators are based on Pawlak's rough set theory, but they can be generalized in several ways. One of the interesting generalizations is to define knowledge operators relative with respect to arbitrary binary relations by using the generalized notion of approximation of a set; see Orlowska [39].

Let R be a binary relation in a set U. For $x \in U$, we define a neighbourhood of x with respect to R:

$$n_R(x) = \{y \in U \mid (x, y) \in R \text{ or } (y, x) \in R\}$$

Then, by lower (upper) approximation of a set $X \subseteq U$, we mean the union of those neighbourhoods which are included in X (which have an element in common with X).

To define the respective knowledge operators, we assume that with every set A of agents there is associated a relation $R(A)$ in the set U. The corresponding *epistemic structure* is $(U, AGT, \{R(A)\}_{A \subseteq AGT}, \{K(A)\}_{A \subseteq AGT})$, where $R : P(AGT) \to P(U \times U)$ assigns binary relations to sets of agents, and $K : P(AGT) \to P(U)$ is an operator such that $K(A)(X)$ is the union of the lower approximation of X and the complement of the upper approximations of X with respect to $R(A)$.

Thus, Kripke structures can be generalized with accessibility relations determined by sets of parameters as:

(KR) $\mathbf{K} = (W, PAR, RL, \{R(P)\}_{P \subseteq PAR, R \in REL})$

where W is a non-empty set of worlds (or states, objects, etc.), PAR is a non-empty set whose elements are called parameters, elements of set REL are mapping $R : P(PAR) \to P(W \times W)$ which assign binary relation in set W to subseteq of set PAR.

Moreover, we assume that $R(P)$ satisfy the following conditions:

$$R(\emptyset) = W \times W$$
$$R(P \cup Q) = R(P) \cap R(Q)$$

The first condition says that the empty set of parameters does not enable us to distinguish any worlds. The second condition says that if we have more parameters then the relation is smaller, less worlds will be glue together.

We here observe that the axiomatization of logics based on Kripke models with relative accessibility relations is an open problem.

2.6 Logics for Knowledge Representation

Although Orlowska's logic is concerned with reasoning about knowledge, we can find several (modal) logics for knowledge representation in the literature, some of which are closely related to rough set theory.

In this section, we review these logics. The starting point of such approaches is an *information system* introduced by Pawlak [46]. An information system is a collection of pieces of information which have the form: object, a list of properties of the object.

Object is anything that can be used in a subject position of a natural language sentence. Although object can be composed and structured in information systems, they are treated as indivisible wholes. Properties of objects are expressed by attributes and their values.

By formal terms, an information system is determined by specifying a non-empty set OB of objects, a non-empty set AT of attributes, a family $\{VAL_a\}_{a \in AT}$ of sets of values of attributes, and an information function f which assigns properties to objects.

There are two types of information functions. *Deterministic* information function is a function of the form $f : OB \times AT \to VAL = \bigcup\{VAL_a \mid a \in AT\}$ which assigns a value of attribute to object. It is assumed that for any $x \in OB$ and $a \in AT$ we have $f(x, a) \in VAL_a$.

Functions of this type determines properties of objects in deterministic way, namely property is uniquely assigned to object.

The other type of information function is *non-deterministic* information function of the form $f : OB \times AT \to P(VAL)$, which assigns a subset of the set of values of attribute to object.

Non-deterministic information function reflects incompleteness of information about properties of objects. The function says what is the range of possible values of every attribute for an object, but the value itself is not known.

An information system is defined as a structure of the form:

$$\mathbf{S} = (OB, AT, \{VAL_a\}_{a \in AT}, f)$$

If f is a deterministic information function, then system $\mathbf{S}$ is called *deterministic information system*, and if f is non-deterministic information function, then $\mathbf{S}$ is called *non-deterministic information system*, respectively.

Information about properties of objects is a basic explicit information included in information systems. From that information, we can derive some other information which is usually expressed in terms of binary relations in the set of objects.

Let $A \subseteq AT$ be a set of attributes. By an *indiscernibility relation* determined by a set A, we mean the relation $ind(A) \subseteq OB \times OB$ defined as:

$$(x, y) \in ind(A) \text{ iff } f(x, a) = f(y, a) \text{ for all } a \in AT$$

For the empty set of attributes, we assume $ind(\emptyset) = OB \times OB$. Thus, two objects stand in relation $ind(A)$ whenever they cannot be distinguished one from the other by means of properties determined by the attribute from the set A.

The following proposition states the basic properties of indiscernibility relations:

Proposition 2.32 *The following properties hold for ind:*

1. *$ind(A)$ is reflexive, symmetric and transitive.*
2. *$ind(A \cup B) = ind(A) \cap ind(B)$.*
3. *$(ind(A) \cup ind(B))^* \subseteq ind(A \cap B)$.*
4. *$A \subseteq B$ imples $ind(B) \subseteq ind(A)$.*

Here, (1) says that indiscernibility relations are equivalence relations. Equivalence class of $ind(A)$ consists of those objects which are indistinguishable up to attributes from the set A.

(2) says that discrimination power of the union of sets of attributes is better than that of the parts of the union. Consequently, the algebra $(\{ind(A)\}_{A \subseteq AT}, \cap)$ is a lower semilattice with the zero element $ind(AT)$.

Indiscernibility of objects plays a crucial role in many applications in which definability of the set of objects is important in terms of properties of single objects. Indiscernibility relations can be derived both from deterministic and non-deterministic information systems.

In connection with non-deterministic information systems, several other relations have been discussed; see Orlowska and Pawlak [43].

Let **S** be a non-deterministic information system. We define a family of *similarity relations* for $A \subseteq AT$, denoted $sim(A)$:

$(x, y) \in sim(A)$ iff $f(x, a) \cap f(y, a) \neq \emptyset$ for all $a \in A$.

We can also consider weak similarity of objects:

$(x, y) \in wsim(A)$ iff there is $a \in A$ such that $f(x, a) \cap f(y, a) \neq \emptyset$.

For some applications, negative similarity might be interesting:

$(x, y) \in nsim(A)$ iff $-f(x, a) \cap -f(y, a) \neq \emptyset$ for all $a \in A$.

Information inclusion of objects $(in(A))$ and weak information inclusion $(win(A))$ are defined as follows:

$(x, y) \in in(A)$ iff $f(x, a) \subseteq f(y, a)$ for all $a \in A$
$(x, y) \in win(A)$ iff there is $a \in A$ such that $f(x, a) \subseteq f(y, a)$.

Observe that the given relations are not independent, they satisfy the following conditions:

Proposition 2.33 *The following conditions hold:*

1. $(x, y) \in in(A)$ and $(x, z) \in sim(A)$ imply $(y, z) \in sim(A)$.
2. $(x, y) \in ind(A)$ implies $(x, y) \in in(A)$.
3. $(x, y) \in in(A)$ and $(y, x) \in in(A)$ imply $(x, y) \in ind(A)$.

It is natural to connect information systems and modal logics. A Kripke structure is of the form $\mathbf{K} = (W, R)$, where W is a non-empty set of possible worlds or states, and $R \subseteq W \times W$ is a binary relation called a accessibility relation.

With every information system of the form **(S)**, we can associate a corresponding Kripke structure $\mathbf{K(S)}$, where set OB is considered to be the universe and relations determined by the system is considered to be accessibility relations (cf. [42]).

Accessibility relations for indiscernibility are assumed to be equivalence relations. Similarity relations are reflexive and symmetric, and information inclusion is reflexive and transitive.

However, we have to assume that a family of indiscernibility relations is closed under intersection, that conditions (1)–(3) in Proposition 2.33 are satisfied in $\mathbf{K(S)}$, providing relationships between indiscernibility, similarity and information inclusion.

Several modal logics for reasoning about objects in information systems have been proposed. Fariñas del Cerro and Orlowska [13] developed data analysis logic *DAL*. The logic can handle inferences in the presence of incomplete information.

Data analysis can be understood as a process of obtaining patterns in a set of data items. They considered two main tasks involved in data analysis in *DAL*, namely

(1) to aggregate data into sets according to their properties,
(2) to define properties adequate for characterization of sets of data.

Obviously, these two tasks are necessary for data analysis.

DAL defined formal counterparts of sets of data and properties. Namely, sets of data are defined by means of the language of *DAL* and properties are defined by means of relational expressions. Thus, data are identified with a non-empty set of objects and a family of equivalence relations on this set.

Objects will be interpreted as data items and relations correspond to properties of data items. Each property induces an equivalence relation such that an equivalence class of the relation consists of those objects which are the same with respect to this property.

The language of *DAL* includes modal operators interpreted as operations of lower and upper approximations determined by indiscernibility relations. Semantics of *DAL* is given by Kripke structures with indiscernibility relations. An algebraic structure is assumed in the set of indiscernibility relations, namely the set is closed under intersection and transitive closure of union of relations.

Expressions of *DAL* are built with symbols from the following pairwise disjoint sets:

VARPROP: a denumerable set of propositional variables
$\{IND_i\}$: a denumerable set of relational constants (i is a natural number)
$\{\cap, \cup^*\}$: the set of relational operations of intersection and transitive closure of union
$\{\neg, \vee, \wedge, \rightarrow\}$: the set of classical propositional operations of negation, disjunction and implication
$\{[\ ], \langle\ \rangle\}$: the set of modal operators.

The set *EREL* of relational expressions is the smallest set satisfying the following conditions:

CONREL $\subseteq$ *EREL*
$IND_1, IND_2 \in EREL$ implies $IND_1 \cap IND_2, IND_1 \cup^* IND_2 \in EREL$

The set *FOR* of formulas is the smallest set such that:

VARPROP $\subseteq$ *FOR*
$F, G \in FOR$ implies $\neg F, F \vee G, F \wedge G, F \rightarrow G \in FOR$
$F \in FOR, IND \in EREL$ imply $[IND]F, \langle IND\rangle F \in FOR$.

A semantics for *DAL* is given by the Kripke model of the form:

(MD) $M = (OB, \{S_p\}_{p \in VARPROP}, \{ind_i\}, m)$

where *OB* is a non-empty set of objects, for any propositional variable p set S_p is a subset of *OB*, for any natural number i ind_i is an equivalence relation in*OB*, and m is a meaning function satisfying the following conditions:

(m1) $m(p) = S_p$ for $p \in VARPOP$, $m(IND_i) = ind_i$ for any i
(m2) $m(IND_1 \cap IND_2) = m(IND_1) \cap m(IND_2)$
(m3) $m(IND_1 \cup^* IND_2) = m(IND_1) \cup^* m(IND_2)$
(m4) $m(\neg F) = -m(F)$
(m5) $m(F \vee G) = m(F) \vee m(G)$
(m6) $m(F \wedge G) = m(F) \wedge m(G)$
(m7) $m(F \to G) = m(F) \to m(G)$
(m8) $m([IND]F) = \{x \in OB \mid$ for all $y \in OB$ if $(x, y) \in m(IND)$ then $y \in m(F))$
(m9) $m(\langle IND \rangle F) = \{x \in OB \mid$ there is $y \in OB$ such that $(x, y) \in m(IND)$ and $y \in m(F))$

A formula is true in model M ($\models_M F$) iff $m(F) = OB$. A formula F is valid ($\models F$) iff F is true in all models.

In the semantics for *DAL* above, formulas are interpreted as subsets of the set of objects from a model, classical propositional operations are the counterparts of set-theoretical operations, and the modal operators $[IND]$ and $\langle IND \rangle]$ correspond to operations of lower and upper approximation, respectively, with respect to indiscernibility relation $m(IND)$.

Proposition 2.34 *The following hold:*

(1) $\models_M F \to [IND]F$ *iff $m(F)$ is $m(IND) - definable$*
(2) $\models_M [IND]F$ *iff $POS(m(IND))m(F) = OB$*
(3) $\models_M \neg[IND]F$ *iff $POS(m(IND))m(F) = \emptyset$*
(4) $\models_M \neg\langle IND \rangle F$ *iff $NEG(m(IND))m(F) = OB$*
(5) $\models_M \langle IND \rangle F$ *iff $NEG(m(IND))m(F) = \emptyset$*
(6) $\models_M \langle IND \rangle F \wedge \langle IND \rangle \neg F$ *iff $BOR(m(IND))m(F) = OB$*
(7) $\models_M \langle IND \rangle F \wedge [IND]\neg F$ *iff $BOR(m(IND))m(F) = \emptyset$*

Since indiscernibility relations, their intersections and transitive closures of union are equivalence relations, operators $[IND]$ and $\langle IND \rangle$ are S5 operators of necessity and possibility, respectively. Hence, all the formulas which are substitutions of theorems of S5 are valid in *DAL*.

Proposition 2.35 *The following hold.*

1. $\models [IND_1]F \vee [IND_2]F \rightarrow [IND_1 \cap IND_2]F$
2. $\models [IND_1 \cup^* IND_2]F \rightarrow [IND_1]F \wedge [IND_2]F$

Note here that it is an open problem of finding a Hilbert style axiomatization of *DAL*.

Model of the form (MD) presented above is said to be a model with *local agreement* of indiscernibility relation whenever for any relations ind_1 and ind_2, and for any object x, the equivalence class of ind_1 determined by x is included in the equivalence class of ind_2 determined by x, or conversely, the class of ind_2 is included in the respective class of ind_1.

The complete axiomatization of *DAL* with respect to the class of models with local agreement is the following:

(D1) All formulas having the form of classical propositional tautology
(D2) $[IND](F \rightarrow G) \rightarrow ([IND]F \rightarrow [IND]G)$
(D3) $[IND]F \rightarrow F$
(D4) $\langle IND \rangle F \rightarrow [IND]\langle IND \rangle F$
(D5) $[IND \cup^* IND_2]F \rightarrow [IND_1]F \wedge [IND_2]F$
(D6) $(([IND_1]F \rightarrow [IND_2]F) \wedge ([IND_1]F \rightarrow [IND_2]F)) \rightarrow ([IND_2]F \rightarrow [IND_2 \cup^* IND_3]F)$
(D7) $[IND_1]F \vee [IND_2]F \rightarrow (IND_1 \cap IND_2]F$
(D8) $(([IND_1]F \rightarrow [IND_3]F) \wedge ([IND_2]F \rightarrow [IND_3]F)) \rightarrow ([IND_1 \cap IND_2]F \rightarrow [IND_3]F)$.

The rules of inference are *modus pones* and necessitation for all operators $[IND]$, i.e., $A/[IND]A$. Axioms (D1)–(D4) are the standard axioms of modal logic S5.

Fariñas del Cerro and Orlowska [13] proved completeness of *DAL*. *DAL* can be extended in various ways. Such extensions involve other operations on relations. For example, we can add the composition of relations. Another possibility is to assume relations which are not necessarily equivalence relations.

Balbiani [6] solved the axiomatization problem of *DAL* and proposed $DAL^\cup$ with a completeness result. He used a Kripke semantics with $\cup$-relative accessibility relation for $DAL^\cup$. See Balbiani [6] for details.

To reason about non-deterministic information logic *NIL* (non-deterministic information logic) has been introduced with modal operators determined by similarity and information inclusion of objects; see Orlowska and Pawlak [43].

The set of formulas of the language of logic *NIL* is the least set including a set *VARPOP* of propositional variables and closed under classical propositional operations and modal operations $[SIM]$, $\langle SIM \rangle$, $[IN]$, $\langle IN \rangle$, $[IN^{-1}]$, $\langle IN^{-1} \rangle$, where *SIM* and *IN* are relational constants interpreted as similarity and informational inclusion and IN^{-1} denotes the converse of relation *IN*.

Kripke models of logic *NIL* are systems pf the form]

(MN) $M = (OB, \{S_p\}_{p \in VARPOP}, \{sim, in\}, m)$

where *OB* is a non-empty set of objects, $S_p \subseteq OB$ for every $p \in VARPROP$, *sim* is a reflexive and transitive relation in *OB*, and moreover, we assume that the relations satisfy the condition:

(n1) If $(x, y) \in in$ and $(x, z) \in sim$, then $(y, z) \in sim$

Meaning function *m* is defined as follows: $m(p) = S_p$ for $p \in VARPROP$, $m(SIM) = sim$, $m(IN) = in$, $m(IN^{-1}) = m(IN)^{-1}$; for complex formulas built with classical propositional operations *m* is given by conditions (m4), ..., (m9), where *IND* is replaced by SIM, IN, IN^{-1}, respectively.

Set of axioms of logic *NIL* consists of axioms of logic B for formulas with operators $[SIM]$ and $\langle SIM \rangle$, axioms of logic S4 for formulas with operators $[IN]$, $\langle IN \rangle$, $[IN^{-1}]$, $\langle$ and the axiom corresponding to condition (n1):

(N1) $[SIM]F \rightarrow [IN^{-1}][SIM][IN]F$

Rules of inference are *modus ponens* and necessitation for the three modal operators of the language.

Vakarelov [55] proved that models of *NIL* represent adequately non-deterministic information systems, as the following proposition:

Proposition 2.36 *For any model of the form (MN) there is a non-deterministic information system of the form (S) with the same set OB of objects such that for any $x, y \in OB$ we have:*

$(x, y) \in sim$ iff $f(x, a) \cap f(y, a) \neq \emptyset$ *for all* $a \in AT$
$(x, y) \in sim$ iff $f(x, a) \subseteq f(y, a)$ *for all* $a \in AT$

Vakarelov [57] introduced information logic *IL* as an extension of *NIL*. The language of *IL* includes the operators of *NIL* and $[IND]$ and $\langle IND \rangle$ determined by indiscernibility relation.

In the corresponding Kripke model, it is assumed that indiscernibility relation *ind* is an equivalence relation and relations *sim*, *in*, *ind* satisfy condition (n1) and the following conditions:

(n2) $ind \subseteq in$
(n3) $in \cap in^{-1} \subseteq ind$

Axioms of *IL* include axioms of *NIL*, axioms of S5 for $[IND]$ and $\langle IND \rangle$, and the following axiom corresponding to (n2):

$[IN]F \rightarrow [IND]F$.

Note that condition (n3) is not expressible by means of a formula of *IL*.

Vakarelov proved completeness of the given set of axioms for the class of models satisfying conditions (n1) and (n2) and the class of models satisfying conditions (n1), (n2) and (n3). He also introduced some other information logics which correspond both to deterministic and non-deterministic information systems.

Logic *NIL* can reason about objects which are incompletely defined. Here, We understand incompleteness as lack of definite information about values of attributes for objects. By using modal operators, we can compare objects with respect to informational inclusion and similarity.

In logic *IL*, it is possible to deal with two kinds of incomplete information: objects are defined up to indiscernibility, and their properties are specified non-deterministically.

Vakarelov also investigated a duality between Pawlak's knowledge representation system and certain information systems of logical type, called *bi-consequence systems*. He developed a complete modal logic *INF* for some informational relations; see Vakarelov [58].

Konikowska [26] proposed a modal logic for reasoning about relative similarity based on the idea of rough set theory. She presented a Kripke semantics and a Gentzen system and proved a completeness result.

As discussed in this section, logics for data analysis and knowledge representation are also of interest to describe incompleteness of knowledge based on the concept of indiscernibility. We can point out that they are more powerful than current knowledge representation languages used in AI.

However, on the theoretical side, the problem of axiomatization remains open for many such logics. The lack of practical proof methods is serious, and detailed studies are needed.

2.7 Fuzzy Logic

Fuzzy logic also formalize vagueness and has many applications. Here, we give a short exposition of fuzzy logic. *Fuzzy set* was proposed by Zadeh [62] to model fuzzy concepts, which cannot be formalized classical set theory. Zadeh also developed a *theory of possibility* based on fuzzy set theory in Zadeh [63]. In fact, fuzzy set theory found many applications in various areas.

Let $\mathcal{U}$ be a set. Then, a fuzzy set is defined as follows:

Definition 2.2 A *fuzzy set* of $\mathcal{U}$ is a function $u : \mathcal{U} \to [0, 1]$. $\mathcal{F}_{\mathcal{U}}$ will denote the set of all fuzzy sets of $\mathcal{U}$.

Several operations on fuzzy sets are defined as follows:

Definition 2.3 For all $u, v \in \mathcal{F}_{\mathcal{U}}$ and $x \in \mathcal{U}$, we put

(1) $(u \vee v)(x) = \sup\{u(x), v(x)\}$
(2) $(u \wedge v)(x) = \inf\{u(x), v(x)\}$
(3) $\bar{u}(x) = 1 - u(x)$.

Definition 2.4 Two fuzzy sets $u, v \in \mathcal{F}_{\mathcal{U}}$ are said to be *equal* iff for every $x \in \mathcal{U}$, $u(x) = v(x)$.

Definition 2.5 $\mathbf{1}_{\mathcal{U}}$ and $\mathbf{0}_{\mathcal{U}}$ are the fuzzy sets of $\mathcal{U}$ such that for all $x \in \mathcal{U}$, $\mathbf{1}_{\mathcal{U}} = 1$ and $\mathbf{0}_{\mathcal{U}} = 0$.

It is easy to prove that $\langle \mathcal{F}_{\mathcal{U}}, \wedge, \vee \rangle$ is a complete lattice having infinite distributive property. Furthermore, $\langle \mathcal{F}_{\mathcal{U}}, \wedge, \vee, ^{-} \rangle$ constitutes an algebra, which in general is not Boolean. Consult Negoita and Ralescu [35] for basics and applications of fuzzy logic.

Since both rough set theory and fuzzy set theory aim to formalize related notions, it is natural to integrate these two theories. In 1990, Dubois and Prade introduced *fuzzy rough sets* as a fuzzy generalization of rough sets [10].

They considered two types of generalizations. One is the upper and lower approximation of a fuzzy set, i.e., *rough fuzzy set*. The other provided an idea of turning the equivalence relation into a fuzzy similarity relation, yielding a *fuzzy rough set*.

Nakamura and Gao [34] also studied fuzzy rough sets and developed a logic for fuzzy data analysis. Their logic can be interpreted as a modal logic based on fuzzy relations. They related similarity relation on a set of objects to rough sets.

2.8   Applications of Rough Set Theory

Now, rough set theory finds many applications. Here, we review some of them. To our best knowledge, it is applied the following areas.

- Mathematics
- Machine Learning
- Data Mining
- Big Data
- Decision Making
- Image Processing
- Switching Circuit
- Robotics
- Medicine

Mathematics could be generalized by using rough set theory. Exploring this line may produce interesting mathematical results, since many mathematical concepts are in fact rough. Mathematics based on rough sets should also be compared to fuzzy mathematics.

Machine Learning (ML) has been mainly studied in AI, giving computers the ability to learn. From the beginning, AI has been concerned with Machinin Learning as its subarea: see Nilsson [37]. Orlowska [40] used modal logic to investigate logical aspects of learning concepts.

A rough set approach to learning was explored by Pawlak [48] who discussed learning from examples and inductive learning. Some of the rough set based applications of Machine Learning involve analysis of the attributes in question and the formulation of the discovered knowledge; see Moshkov and Zielosko [29].

Data Mining (DM) studies the process of finding patterns in large databases. It is also called *knowledge discovery in database* (KDD). For Data Mining, consult Adriaans and Zatinge [1] for details. The main goal of Data Mining is to extract meaningful knowledge from a large database. In this regard, rough set theory is useful for extraction of decision rules from databases. For the subject, consult Lin and Cercone [28].

Big Data is the area which deals with a large and complex data collection. The term sometimes refers to such data themselves. The treatment of Big Data is beyond traditional database technology and requires new approaches. Note that Big Data is also related to Data Mining.

Rough set theory can properly offer interesting approaches to Big Data including parallel algorithms to compute equivalence classes, decision classes, approximations, etc. For instance, the approaches explored in Pradeepa and Selvadoss ThnamaniLee [50] and Zhang et al. [64] may be relevant to Big Data.

Decision making is the process of selecting a logical choice from possible options. A system capable of doing decision making is called a *decision support system*, and it is of special use for various human decision making.

There are several advantages of the use of rough set theory for decision making. We do not need any preliminary information about data. We have efficient methods for finding hidden patterns in data. We can evaluate the significance of data and easily understand. For example, Decision table and its simplification methods can serve as theoretical foundations for decision-making; see Slowinski et al. [53].

Image processing studies image data and their various treatments. Indeed various techniques for imaging processing has been extensively investigated, but rough set theory can provide alternative techniques.

For example, it is helpful to segmentation and extraction; see Pal et al. [45]. Another promising application is image classification and retrieval, which challenge standard approaches. Note that rough set theory can be also applied to the so-called *pattern recognition*.

Switching circuit is a basis for hardware design and some effective methods like *Karnaugh maps* have been already established (cf. [25]). However, it is very important to improve the efficiency, accuracy and low power consumption in an electronic circuit, and we need circuit minimization.

Rough set theory can provide an alternative base for switching circuit. Since switching function can be described as a decision table, we can employ simplification methods for it. Pawlak gave an approach to switching circuit; see Pawlak [48] for details.

Robotics is the area of constructing a robot. There are several types of robotics systems; from simple one to human-like one. In any case, it needs various disciplines mainly for hardware and software. It is observed that any robot should handle rough information. For an overview on the subject, the reader is referred to Bit and Beaubouef [7].

Medicine is one of the most significant areas for rough set theory, since medical data can be considered both incomplete and vague. However, physicians must diagnose a patient and decide the best way for a patient without obtaining complete information. We can find many approaches to medical science based on rough set theory.

For instance, Tsumoto [54] proposed a model for medical diagnostic rules based on rough sets. Hirano and Tsumoto [21] applied rough set theory to the analysis of medical images.

References

1. Adriaans, P., Zantinge, D.: Data Mining. Addison-Wesley, Reading, MA (1996)
2. Akama, S., Murai, T., Kudo, Y.: Da Costa logics and vagueness. In: Proceedings of GrC2014, Noboribetsu, Japan (2014)
3. Akama, S., Murai, T., Kudo, Y.: Heyting-Brouwer rough set logic. In: Proceedings of KSE2013, Hanoi, pp. 135–145. Springer, Heidelberg (2013)
4. Akama, S., Murai, T., Kudo, Y.: Reasoning with Rough Sets. Springer, Heidelberg (2018)
5. Armstrong, W.: Dependency structures in data base relationships. In: IFIP'74, pp. 580–583 (1974)
6. Balbiani, P.: A modal logic for data analysis. In: Proceedings of MFCS'96. LNCS, vol. 1113, pp. 167–179. Springer, Berlin
7. Bit, M., Beaubouef, T.: Rough set uncertainty for robotic systems. J. Comput. Syst. Colleges **23**, 126–132 (2008)
8. Chellas, B.: Modal Logic: An Introduction. Cambridge University Press, Cambridge (1980)
9. de Caro, F.: Graded modalities II. Stud. Logica **47**, 1–10 (1988)
10. Dubois, D., Prade, H.: Rough fuzzy sets and fuzzy rough sets. Int. J. Gen. Syst. **17**, 191–209 (1989)
11. Düntsch, I.: A logic for rough sets. Theor. Comput. Sci. **179**, 427–436 (1997)
12. Fagin, R., Halpern, J., Moses, Y., Vardi, M.: Reasoning about Knowledge. MIT Press, Cambridge, MA (1995)
13. Fariñas del Cerro, L., Orlowska, E.: DAL—a logic for data analysis. Theor. Comput. Sci. **36**, 251–264 (1985)
14. Fattorosi-Barnaba, M., Amati, G.: Modal operators with probabilistic interpretations I. Stud. Logica **46**, 383–393 (1987)
15. Fattorosi-Barnaba, M., de Caro, F.: Graded modalities I. Stud. Logica **44**, 197–221 (1985)
16. Fattorosi-Barnaba, M., de Caro, F.: Graded modalities III. Stud. Logica **47**, 99–110 (1988)
17. Ganter, B., Wille, R.: Formal Concept Analysis. Springer, Berlin (1999)
18. Halpern, J., Moses, Y.: Towards a theory of knowledge and ignorance: preliminary report. In: Apt, K. (ed.) Logics and Models of Concurrent Systems, pp. 459–476. Springer, Berlin (1985)
19. Halpern, J., Moses, Y.: A theory of knowledge and ignorance for many agents. J. Logic Comput. **7**, 79–108 (1997)
20. Hintikka, S.: Knowledge and Belief. Cornell University Press, Ithaca (1962)
21. Hirano, S., Tsumoto, S.: Rough representation of a region of interest in medical images. Int. J. Approx. Reason. **40**, 23–34 (2005)
22. Iturrioz, L.: Rough sets and three-valued structures. In: Orlowska, E. (ed.) Logic at Work: Essays Dedicated to the Memory of Helena Rasiowa, pp. 596–603. Physica-Verlag, Heidelberg (1999)
23. Iwinski, T.: Algebraic approach to rough sets. Bull. Pol. Acad. Math. **37**, 673–683 (1987)
24. Järvinen, J., Pagliani, P., Radeleczki, S.: Information completeness in Nelson algebras of rough sets induced by quasiorders. Stud. Logica **101**, 1073–1092 (2013)
25. Karnaugh, M.: A map method for synthesis of combinattional logic circuits. Trans. AIEE Commun. Electoron. **72**, 593–599 (1953)
26. Konikowska, B.: A logic for reasoning about relative similarity. Stud. Logica **58**, 185–228 (1997)
27. Liau, C.-J.: An overview of rough set semantics for modal and quantifier logics. Int. J. Uncertain. Fuzziness Knowl.-Based Syst. **8**, 93–118 (2000)

28. Lin, T.Y., Cercone, N.: Rough Stes and Data Mining. Springer, Heidelberg (1997)
29. Moshkov, M., Zieloska, B.: Combinatorial Machine Learning: A Rough Set Approach. Springer, Heidelberg (2011)
30. Murai, T., Miyakoshi, M., Shimbo, M.: Soundness and completeness theorems between the Dempster-Shafer theory and logic of belief. In: Proceedings of 3rd FUZZ-IEEE (WCCI), pp. 855–858 (1994)
31. Murai, T., Miyakoshi, M., Shimbo, M.: Soundness and completeness theorems between the Dempster-Shafer theory and logic of belief. In: Proceedings of 3rd FUZZ-IEEE (WCCI), pp. 855–858 (1994)
32. Murai, T., Miyakoshi, M., Shinmbo, M.: Measure-based semantics for modal logic. In: Lowen, R., Roubens, M. (eds.) Fuzzy Logic: State of the Arts, pp. 395–405. Kluwer, Dordrecht (1993)
33. Murai, T., Miyakoshi, M., Shinmbo, M.: Measure-based semantics for modal logic. In: Lowen, R., Roubens, M. (eds.) Fuzzy Logic: State of the Arts, pp. 395–405. Kluwer, Dordrecht (1993)
34. Nakamura, A., Gao, J.: A logic for fuzzy data analysis. Fuzzy Sets Syst. **39**, 127–132 (1991)
35. Negoita, C., Ralescu, D.: Applications of Fuzzy Sets to Systems Analysis. Wiley, New York (1975)
36. Nelson, D.: Constructible falsity. J. Symb. Logic **14**, 16–26 (1949)
37. Nillson, N.: The Mathematical Foundations of Learning Machines. Morgan Kaufmann (1990)
38. Ore, O.: Galois connexion. Trans. Am. Math. Soc. **33**, 493–513 (1944)
39. Orlowska, E.: Kripke models with relative accessibility relations and their applications to inferences from incomplete information. In: Mirkowska, G., Rasiowa, H. (eds.) Mathematical Problems in Computation Theory, pp. 327–337. Polish Scientific Publishers, Warsaw (1987)
40. Orlowska, E.: Logical aspects of learning concepts. Int. J. Approx. Reason. **2**, 349–364 (1988)
41. Orlowska, E.: Logic for reasoning about knowledge. Z. Math. Logik Grundl. Math. **35**, 559–572 (1989)
42. Orlowska, E.: Kripke semantics for knowledge representation logics. Stud. Logica **49**, 255–272 (1990)
43. Orlowska, E., Pawlak, Z.: Representation of nondeterministic information. Theor. Comput. Sci. **29**, 27–39 (1984)
44. Pagliani, P.: Rough sets and Nelson algebras. Fundam. Math. **27**, 205–219 (1996)
45. Pal, K., Shanker, B., Mitra, P.: Granular computing, rough entropy and object extraction. Pattern Recognit. Lett. **26**, 2509–2517 (2005)
46. Pawlak, Z.: Information systems: theoretical foundations. Inf. Syst. **6**, 205–218 (1981)
47. Pawlak, Z.: Rough sets. Int. J. Comput. Inf. Sci. **11**, 341–356 (1982)
48. Pawlak, Z.: Rough Sets: Theoretical Aspects of Reasoning about Data. Kluwer, Dordrecht (1991)
49. Pomykala, J., Pomykala, J.A.: The Stone algebra of rough sets. Bull. Pol. Acad. Sci. Math. **36**, 495–508 (1988)
50. Pradeepa, A., Selvadoss ThanamaniLee, A.: Hadoop file system and fundamental concept of mapreduce interior and closure rough set approximations. Int. J. Adv. Res. Comput. Commun. Eng. **2** (2013)
51. Rasiowa, H.: An Algebraic Approach to Non-classical Logics. North-Holland, Amsterdam (1974)
52. Sendlewski, A.: Nelson algebras through Heyting ones I. Stud. Logica **49**, 105–126 (1990)
53. Slowinski, R., Greco, S., Matarazzo, B.: Rough sets and decision making. In: Meyers, R. (ed.) Encyclopedia of Complexity and Systems Science, pp. 7753–7787. Springer, Heidelberg (2009)
54. Tsumoto, S.: Modelling medical diagnostic rules based on rough sets. Rough Sets and Current Trends in Computing, pp. 475–482 (1998)
55. Vakarelov, D.: Abstract characterization of some knowledge representation systems and the logic *NIL* of nondeterministic information. In: Skordev, D. (ed.) Mathematical Logic and Applications. Plenum Press, New York (1987)
56. Vakarelov, D.: Notes on constructive logic with strong negation. Stud. Logica **36**, 110–125 (1977)

57. Vakarelov, D.: Modal logics for knowledge representation systems. Theor. Comput. Sci. **90**, 433–456 (1991)
58. Vakarelov, D.: A modal logic for similarity relations in Pawlak knowledge representation systems. Stud. Logica **55**, 205–228 (1995)
59. Wong, S., Ziarko, W.: Comparison of the probabilistic approximate classification and the fuzzy set model. Fuzzy Sets Syst. **21**, 357–362 (1987)
60. Yao, Y., Wong, S., Lingras, P.: A decision-theoretic rough set model. In: Ras, Z., Zemankova, M., Emrich, M. (eds.) Methodology for Intelligent Systems, vol. 5, pp. 17–24. North-Holland, New York (1990)
61. Yao, Y., Lin, T.: Generalization of rough sets using modal logics. Intell. Autom. Soft Comput. **2**, 103–120 (1996)
62. Zadeh, L.: Fuzzy sets. Inf. Control **8**, 338–353 (1965)
63. Zadeh, L.: Fuzzy sets as a basis for a theory of possibility. Fuzzy Sets Syst. **1**, 3–28 (1976)
64. Zhang, J., Li, T., Ruan, D., Gao, Z., Zhao, C.: A parallel method for computing rough set approximations. Inf. Sci. **194**, 209–223 (2012)
65. Ziarko, W.: Variable precision rough set model. J. Comput. Syst. Sci. **46**, 39–59 (1993)

Chapter 3
Object Reduction in Rough Set Theory

Abstract This chapter deals with object reduction in rough set theory. We introduce a concept of object reduction that reduces the number of objects as long as possible with keeping the results of attribute reduction in the original decision table.

Keywords Rough set · Object reduction · Attribute reduction · Discernibility matrix

3.1 Introduction

Attribute reduction is one of the most popular research topics in the community of *rough set theory*. In Pawlak's rough set theory [2], attribute reduction computes minimal subsets of condition attributes in a given decision table that keep the classification ability by the condition attributes. Such minimal subsets of condition attributes are called *relative reducts*, and a computation method of all relative reducts using the discernibility matrix has been proposed in [3]. On the other hand, to the best of our knowledge, reduction of objects in some sense is not a popular topic in rough set theory.

In this chapter, from a viewpoint of computing all relative reducts, we introduce a concept of object reduction that reduces the number of objects as long as possible with keeping the results of attribute reduction in the original decision table.

3.2 Rough Sets

In this section, we briefly review Pawlak's rough set theory. Contents of this section is based on [1].

© Springer Nature Switzerland AG 2020
S. Akama et al., *Topics in Rough Set Theory*, Intelligent Systems
Reference Library 168, https://doi.org/10.1007/978-3-030-29566-0_3

3.2.1 Decision Tables, Indiscernibility Relations, and Lower Approximations

Let C be a finite set of condition attributes, and $\mathsf{d} \notin C$ be a decision attribute. The following structure is called a *decision table* to represent a table-style dataset as the target of rough set-based data analysis:

$$DT = (U, C \cup \{\mathsf{d}\}). \tag{3.1}$$

Each attribute $\mathsf{a} \in C \cup \{d\}$ is a function $\mathsf{a} : U \to V_{\mathsf{a}}$, where V_{a} is a finite set of values of the attribute a. It is well known that equivalence relations defined on the set U provides partitions of U.

Each subset $B \subseteq C \cup \{\mathsf{d}\}$ of attributes constructs an equivalence relation on U, called an *indiscernibility relation* with respect to B, as follows:

$$IND(B) = \{(x, y) \mid \mathsf{a}(x) = \mathsf{a}(y), \forall \mathsf{a} \in B\}. \tag{3.2}$$

From the indiscernibility relation $IND(B)$ and an object $x \in U$, an equivalence class $[x]_B$ is obtained.

As we mentioned, the set of all equivalence classes with respect to the indiscernibility relation $IND(B)$, called the quotient set, $U/IND(B) = \{[x]_B \mid x \in U\}$ is a partition of U. Particularly, the quotient set by the indiscernibility relation $IND(\{\mathsf{d}\})$ with respect to the decision attribute d is called the set of decision classes and denoted by $\mathcal{D} = \{D_1, \ldots, D_m\}$.

For considering attribute reduction in the given decision table, we introduce the lower approximation for each decision class D_i $(1 \le i \le m)$ by a subset $B \subseteq C$ of condition attributes as follows:

$$\underline{B}(D_i) = \{x \in U \mid [x]_B \subseteq D_i\}. \tag{3.3}$$

The lower approximation $\underline{B}(D_i)$ is the set of objects that are correctly classified to D_i by using the information of B.

3.2.2 Relative Reducts

From the viewpoint of classification of objects, minimal subsets of condition attributes for classifying all discernible objects to correct decision classes are convenient. Such minimal subsets of condition attributes are called *relative reducts* of the given decision table. To formally define the relative reducts, we introduce the concept of positive region. Let $B \subseteq C$ be a set of condition attributes. The positive region of the partition $\mathcal{D}$ by B is defined by

$$POS_B(\mathcal{D}) = \bigcup_{D_i \in \mathcal{D}} \underline{B}(D_i). \tag{3.4}$$

The positive region $POS_B(\mathcal{D})$ is the set of objects classified to correct decision classes by checking the attribute values in every attribute in B. Particularly, the set $POS_C(\mathcal{D})$ is the set of all discernible objects in the decision table DT.

Here, we define the relative reducts. A set $A \subseteq C$ is called a relative reduct of the decision table DT if the set A satisfies the following two conditions:

1. $POS_A(\mathcal{D}) = POS_C(\mathcal{D})$.
2. $POS_B(\mathcal{D}) \neq POS_C(\mathcal{D})$ for any proper subset $B \subset A$.

In general, there are plural relative reducts in a decision table. For a given decision table DT, we denote the set of all relative reducts of DT by $RED(DT)$.

3.2.3 Discernibility Matrix

Discernibility matrix was firstly introduced by Skowron and Rauszer [3] to extract all relative reducts from a decision table. Suppose that the set of objects U in the decision table DT has n objects. The discernibility matrix DM of DT is a symmetric $n \times n$ matrix and its element at the ith row and jth column in DM is the following set of condition attributes:

$$\delta_{ij} = \begin{cases} \{\mathsf{a} \in C \mid \mathsf{a}(x_i) \neq \mathsf{a}(x_j)\}, \\ \qquad \text{if } \mathsf{d}(x_i) \neq \mathsf{d}(x_j) \text{ and} \\ \qquad \{x_i, x_j\} \cap POS_C(\mathcal{D}) \neq \emptyset, \\ \emptyset, \qquad\qquad\qquad\qquad \text{otherwise.} \end{cases} \tag{3.5}$$

The element δ_{ij} means that the objects x_i is discernible from the object x_j by comparing at least one attribute $\mathsf{a} \in \delta_{ij}$.

Let $\delta_{ij} = \{\mathsf{a}_1, \dots, \mathsf{a}_l\}$ be the element of DT at ith row and j-column. Contents of each element δ_{ij} is represented by the logical formula as follows:

$$L(\delta_{ij}) : \mathsf{a}_1 \vee \cdots \vee \mathsf{a}_l. \tag{3.6}$$

By constructing the conjunctive normal form from the logical formulas and transforming the formula to the prime implicant, all relative reducts in the decision table are computed. The problem of extracting all relative reducts from the given decision table is, however, an NP-hard problem [3], which concludes that computation of all relative reducts from a decision table with numerous objects and attributes is intractable.

Table 3.1 An example of a decision table

	c_1	c_2	c_3	c_4	c_5	c_6	d
x_1	1	0	0	0	0	1	1
x_2	0	1	0	0	0	1	1
x_3	0	2	1	0	1	0	2
x_4	0	1	1	1	1	0	2
x_5	0	1	2	0	0	1	1
x_6	1	1	0	0	1	1	3

Table 3.2 The discernibility matrix of Table 3.1

	x_1	x_2	x_3	x_4	x_5	x_6
x_1	$\emptyset$					
x_2	$\emptyset$	$\emptyset$				
x_3	$\begin{Bmatrix} c_1, c_2, \\ c_3, c_5, \\ c_6 \end{Bmatrix}$	$\begin{Bmatrix} c_2, c_3, \\ c_5, c_6 \end{Bmatrix}$	$\emptyset$			
x_4	$\begin{Bmatrix} c_1, c_2, \\ c_3, c_4, \\ c_5, c_6 \end{Bmatrix}$	$\begin{Bmatrix} c_3, c_4, \\ c_5, c_6 \end{Bmatrix}$	$\emptyset$	$\emptyset$		
x_5	$\emptyset$	$\emptyset$	$\begin{Bmatrix} c_2, c_3, \\ c_5, c_6 \end{Bmatrix}$	$\begin{Bmatrix} c_3, c_4, \\ c_5, c_6 \end{Bmatrix}$	$\emptyset$	
x_6	$\{c_2, c_5\}$	$\{c_1, c_5\}$	$\begin{Bmatrix} c_1, c_2, \\ c_3, c_6 \end{Bmatrix}$	$\begin{Bmatrix} c_1, c_3, \\ c_4, c_6 \end{Bmatrix}$	$\begin{Bmatrix} c_1, c_3, \\ c_5 \end{Bmatrix}$	$\emptyset$

Example 3.1 Table 3.1 is an example of a decision table with the set of objects $U = \{x_1, \ldots, x_6\}$, the set of condition attributes $C = \{c_1, \ldots, c_6\}$, and $d \notin C$ is the decision attribute.

Table 3.2 shows the discernibility matrix of the decision table in Table 3.1. In Table 3.2, we represent only the lower triangular part of the discernibility matrix. The element $\delta_{61} = \{c_2, c_5\}$ means that the element x_6 with $d(x_6) = 3$ is discernible from x_1 with $d(x_1) = 1$ by comparing either the value of the attribute c_2 or c_5 between x_6 and x_1.

From the discernibility matrix in Table 3.2, for example, a logical formula $L(\delta_{61}) = c_2 \vee c_5$ is constructed based on the nonempty element δ_{61}. By connecting all the logical formulas, the following conjunctive normal form is obtained:

$$(c_1 \vee c_2 \vee c_3 \vee c_5 \vee c_6) \wedge (c_1 \vee c_2 \vee c_3 \vee c_4 \vee c_5 \vee c_6)$$
$$\wedge (c_2 \vee c_5) \wedge (c_2 \vee c_3 \vee c_5 \vee c_6) \wedge (c_3 \vee c_4 \vee c_5 \vee c_6)$$
$$\wedge (c_1 \vee c_5) \wedge (c_2 \vee c_3 \vee c_5 \vee c_6) \wedge (c_1 \vee c_2 \vee c_3 \vee c_6)$$
$$\wedge (c_3 \vee c_4 \vee c_5 \vee c_6) \wedge (c_1 \vee c_3 \vee c_4 \vee c_6) \wedge (c_1 \vee c_3 \vee c_5).$$

This conjunctive normal form has many redundant terms. Such redundant terms are eliminated by using idempotent law $P \wedge P = P$ and absorption law $P \wedge (P \vee Q) = P$, where P and Q are any logical formulas. We then obtain the following simplified conjunctive normal form:

$$(c_2 \vee c_5) \wedge (c_3 \vee c_4 \vee c_5 \vee c_6)$$
$$\wedge(c_1 \vee c_5) \wedge (c_1 \vee c_2 \vee c_3 \vee c_6)$$
$$\wedge(c_1 \vee c_3 \vee c_4 \vee c_6).$$

After repeating the application of distributive law, idempotent law, and absorption law, the following prime implicant is obtained:

$$(c_5 \wedge c_6) \vee (c_3 \wedge c_5) \vee (c_1 \wedge c_5) \vee (c_1 \wedge c_2 \wedge c_3)$$
$$\vee(c_1 \wedge c_2 \wedge c_4) \vee (c_1 \wedge c_2 \wedge c_6) \vee (c_2 \wedge c_4 \wedge c_5).$$

It concludes that there are seven relative reducts in the decision table, i.e., $\{c_5, c_6\}$, $\{c_3, c_5\}$, $\{c_1, c_5\}$, $\{c_1, c_2, c_3\}$, $\{c_1, c_2, c_4\}$, $\{c_1, c_2, c_6\}$, and $\{c_2, c_4, c_5\}$.

3.3 Proposal of Object Reduction

In this section, we propose a concept of object reduction in rough set theory from a viewpoint of computing all relative reducts in a given decision table.

3.3.1 Definition of Object Reduction

As we illustrated in Example 3.1, many elements in the discernibility matrix that corresponds to the logical formulas $L(\delta)$ are redundant for computing relative reducts, and these logical formulas are eliminated by using idempotent law and absorption law. This fact means that many elements in the discernibility matrix may not work for computation of all relative reducts of the given decision table.

This fact also indicates that some objects in the decision table may not work for computation of all relative reducts, and from a viewpoint of attribute reduction, such objects may be reducible from the decision table.

In this chapter, we introduce a concept of object reduction of the given decision table that keeps all relative reducts identical to the original decision table.

Definition 3.1 Let $DT = (U, C \cup \{d\})$ be a given decision table and $RED(DT)$ be the set of all relative reducts of DT. Suppose that $x \in U$ is an object of DT, $DT' = (U \setminus \{x\}, C \cup \{d\})$ be the decision table that the object x is removed from U, where the domain of each attribute $a \in C \cup \{d\}$ is restricted to $U \setminus \{x\}$, i.e., $a : U \setminus \{x\} \rightarrow V_a$, and $RED(DT')$ is the set of all relative reducts of the decision table DT'. The object $x \in U$ is called an irreducible object of DT if and only if $RED(DT') \neq RED(DT)$ holds. The object $x \in U$ that is not an irreducible object of DT is called a possibly reducible object of DT.

By this definition, rejection of an irreducible object from the original decision table DT causes some change of results of attribute reduction from DT.

In general, removing an irreducible object $x_i \in U$ in DT makes some object $x_j \in U$ such that there is need to discern x_j from x_i not discernible from any other objects $x_k \in U$ including x_i, which indicates that the condition attributes in the element δ_{ij} in the original discernibility matrix DM may not appear in relative reducts. On the other hand, rejection of a possibly reducible object does not affect to attribute reduction.

The term "possibly reducible" means that not all possible reducible objects may be reducible from the original decision table, i.e., rejection of a possibly reducible object in the original decision table DT may make some other possibly reducible object in DT an irreducible object in the resulted decision table DT'.

Example 3.2 Here, we consider two examples of decision tables based on the original decision table by Table 3.1. The first example is a decision table by removing the object $x_5 \in U$ from Table 3.1 and we denote this decision table as DT_{x_5}. To consider computing all relative reducts from the decision table DT_{x_5}, we ignore the row and column of x_5 in the discernibility matrix of DT in Table 3.2, applying idempotent law and absorption law to the logical formula constructed from the discernibility matrix, and the following conjunctive normal form is obtained:

$$(c_2 \vee c_5) \wedge (c_3 \vee c_4 \vee c_5 \vee c_6)$$
$$\wedge (c_1 \vee c_5) \wedge (c_1 \vee c_2 \vee c_3 \vee c_6)$$
$$\wedge (c_1 \vee c_3 \vee c_4 \vee c_6).$$

This formula is identical to the conjunctive normal form that we illustrated in Example 3.1, which concludes that all relative reducts obtained from the decision table DT_{x_5} are identical to the relative reducts from the original decision table DT. Therefore, $RED(DT_{x_5}) = RED(DT)$ holds and the object $x_5 \in U$ is a possibly reducible object of DT.

Next example is a decision table by removing the object $x_4 \in U$ from DT and we denote this decision table as DT_{x_4}. Similar to the case of DT_{x_5}, we ignore the row and column of x_4 in Table 3.2, applying idempotent law and absorption law, and the following conjunctive normal form is obtained:

$$(c_2 \vee c_5) \wedge (c_1 \vee c_5) \wedge (c_1 \vee c_2 \vee c_3 \vee c_6).$$

This formula is transformed to the following prime implicant:

$$(c_1 \wedge c_2) \vee (c_1 \wedge c_5) \vee (c_2 \wedge c_5) \vee (c_3 \wedge c_5) \vee (c_5 \wedge c_6),$$

which concludes that there are five relative reduct in DT_{x_4}, i.e., $\{c_1, c_2\}$, $\{c_1, c_5\}$, $\{c_2, c_5\}$, $\{c_3, c_5\}$, and $\{c_5, c_6\}$. Therefore, $RED(DT_{x_4}) \neq RED(DT)$ holds and the object $x_4 \in U$ is an irreducible object of DT.

3.3.2 Properties of Possibly Reducible Objects and Irreducible Objects

Discussion in Example 3.2 indicates close relationship between the concept of reducibility of objects and discernibility matrix. In this subsection, we consider theoretical connections between possibly reducible objects and discernibility matrix.

Proposition 3.1 *Let $DT = (U, C \cup \{d\})$ be a given decision table and DM be the discernibility matrix of DT. An object $x_i \in U$ is a possibly reducible object of DT if and only if, for every nonempty element δ_{ij} $(1 \leq j \leq |U|)$, there exists an element δ_{kl} in DM that either $i \neq k$ or $j \neq l$ and $\delta_{kl} \subseteq \delta_{ij}$ hold.*

This proposition means that, for every nonempty element δ_{ij} that is required to discern the possibly reducible object x_i from another object x_j, the corresponding logical formula $L(\delta_{ij})$ will be eliminated by the logical formula $L(\delta_{kl})$ by using idempotent law or absorption law. Then, information to discern the object x_i from the object x_j is already included in the element δ_{kl} to discern other objects x_k and x_l, and therefore, discerning x_i from x_j has no meaning from the viewpoint of attribute reduction.

Corollary 3.1 *Let DT be a given decision table and DM be the discernibility matrix of DT. An object $x_i \in U$ is an irreducible object of DT if and only if there exists a nonempty element δ_{ik} in DM such that any nonempty element $\delta \neq \emptyset$ in DM is not a proper subset of δ_{ik}, i.e., $\delta \not\subset \delta_{ik}$ holds.*

The original definition of irreducible objects and possibly reducible objects are based on the set of all relative reducts $RED(DT)$ of the given decision table DT. As we mentioned in Sect. 3.2.3, however, computing all relative reducts is actually impossible from the decision table with numerous objects and attributes. Hence, the original concept of reducibility of objects is not computable.

On the other hand, by Proposition 3.1, now we can compute possibly reducible objects concretely. This computation is based on making the discernibility matrix and comparison of elements in the discernibility matrix by set inclusion relationship. Consequently, there is no need to compute all relative reducts to find possibly reducible objects.

3.3.3 An Algorithm for Object Reduction

In this subsection, we introduce an algorithm for computing a result of object reduction by removing possibly reducible objects in a given decision table as many as possible. Relative reducts of the resulted decision table are identical to the ones of the original decision table.

Algorithm 1 outputs a result of object reduction of the given decision table. Steps 3–9 correspond to elimination of redundant elements δ_{ij} in the discernibility matrix DM by using idempotent law and absorption law. Steps 10–14 remove objects x_i as possibly reducible objects because all elements δ_{ij} to discern x_i from another discernible objects x_j are replaced to empty set, i.e., discerning x_i from other objects is redundant for attribute reduction.

Algorithm 1 object reduction algorithm

Input: decision table $DT = (U, C \cup \{d\})$
Output: result of object reduction $DT' = (U', C \cup \{d\})$
1: Compute the discernibility matrix DM of DT
2: $U' \leftarrow U$
3: **for all** $\delta_{kl} \in DM$ **do**
4:     **for all** $\delta_{ij} \in DM$ **do**
5:         **if** ($i \neq k$ **or** $j \neq l$) **and** $\delta_{kl} \subseteq \delta_{ij}$ **then**
6:             $\delta_{ij} \leftarrow \emptyset$
7:         **end if**
8:     **end for**
9: **end for**
10: **for** $i = 1$ to $|U|$ **do**
11:     **if** $\delta_{ij} = \emptyset$ for all $j \in \{1, \ldots, |U|\}$ **then**
12:         $U' \leftarrow U' \setminus \{x_i\}$
13:     **end if**
14: **end for**
15: **return** $DT' = (U', C \cup \{d\})$.

Example 3.3 We show an example of object reduction of Table 3.1 by Algorithm 1. Comparing elements in DM by Table 3.2 and replacing redundant elements δ_{ij} to empty set, as a result, the following nonempty elements are obtained: $\delta_{61} = \{c_2, c_5\}, \delta_{62} = \{c_1, c_5\}, \delta_{42} = \{c_3, c_4, c_5, c_6\}, \delta_{63} = \{c_1, c_2, c_3, c_6\}$, and $\delta_{64} = \{c_1, c_3, c_4, c_6\}$.

Consequently, $\delta_{5j} = \emptyset$ holds for all $j \in \{1, \ldots, 6\}$, the object x_5 is a possibly redundant object and it is removed from U.

As we have shown in Example 3.2, the resulted decision table generates all relative reducts that are identical to the relative reducts in the original decision table.

3.3.4 Application to Dataset

We used Algorithm 1 to Zoo dataset in UCI Machine Learning Repository [4]. The Zoo dataset consists of 101 animals (objects) and 17 attributes like "hair", "feather", "egg", etc. We set the attribute "type" as the decision attribute, and remaining 16 attributes as condition attributes.

The decision attribute "type" divides 101 animals to 7 classes like "mammalian", "birds", "reptiles", "amphibians", "fishes", "insects", and "others". We confirmed that 33 relative reducts are obtained from the original Zoo dataset.

As a result of applying Algorithm 1 to the Zoo dataset, 86 objects were removed as possibly reducible objects. We also confirmed that 33 relative reducts were obtained from the resulted decision table and all relative reducts were identical to the relative reducts from the original Zoo dataset.

3.4 Conclusions

In this chapter, we proposed an approach of object reduction in rough set theory. Our approach is based on removing as many objects as possible with keeping the result of extraction of all relative reducts. We introduced the concept of possibly reducible objects as the objects that do not affect the result of computing all relative reducts by removing the objects from the decision table, and the concept of irreducible objects as the objects that removing the objects changes the result of computing all relative reducts.

We showed theoretical connection between possibly reducible objects and the discernibility matrix of the decision table, and also proposed an algorithm to compute object reduction based on the theoretical connection. Experiment result indicates that the proposed approach can efficiently reduce the number of objects with keeping the ability of attribute reduction.

Future works are more refinement of theoretical connection between possibly reducible objects and discernibility matrix, and further experiments of the proposed approach using various datasets. To determine the remaining element in the discernibility matrix when using the idempotent law and absorption. Moreover, object-wise search technique is required to select the remaining element to remove as many objects as possible. This refinement is an important future issue.

References

1. Mori, N., Tanaka, H., Inoue, K. (eds.): Rough sets and Kansei: knowledge acquisition and reasoning from Kansei data, Kaibundo (2004) (in Japanese)
2. Pawlak, Z.: Rough Sets: Theoretical Aspects of Reasoning About Data. Kluwer, Dordrecht (1991)

3. Skowron, A., Rauszer, C.M.: The discernibility matrix and functions in information systems. In: Słowiński, R. (ed.) Intelligent Decision Support: Handbook of Application and Advance of the Rough Set Theory, pp. 331–362. Kluwer, Dordrecht (1992)
4. UCI machine learning repository. http://archive.ics.uci.edu/ml/index.php

Chapter 4
Recommendation Method by Direct Setting of Preference Patterns Based on Interrelationship Mining

Abstract This chapter introduces a recommendation method by directly setting of users' preference patterns of items for similarity evaluation between users. Yamawaki et al. proposed a recommendation method based on comparing users' preference patterns instead of directly comparing users' rating scores. However, Yamawaki et al.'s method for extraction of preference patterns has limitation of representation ability and this method is not applicable in the case that a user has "cyclic" preference patterns. Our method for extraction of preference patterns is based on partial pairwise comparison of items and it is applicable to represent "cyclic" preference patterns. A tourist spot recommender system is implemented as a prototype of recommender system based on the proposed approach and the implemented recommender system is evaluated by experiments.

Keywords Recommendation method · Preference pattern · Interrelationship mining · Rough set · Tourist spot recommender system

4.1 Introduction

Nowadays, with the rapid growth of the amount of information on internet, recommender systems (RSs) become popular in various aspects of our daily life. There are various approaches of recommender systems, however, difficulty of providing recommendations with both high accuracy and high diversity is pointed out (cf. Bradley and Smyth [1]), and there are many approaches that aims at providing recommendations with high accuracy and high diversity (e.g. [2, 3, 12, 16, 17]). For example, Yamawaki et al. [13, 14] proposed a recommendation method that is based on similarity of preference patterns between the active user and other users instead of directly comparing rating scores in the framework of collaborative filtering.

In this chapter, instead of making active user's preference patterns from rating scores of the active user, we introduce a recommendation method that the active user inputs her/his preference patterns directly.

© Springer Nature Switzerland AG 2020

S. Akama et al., *Topics in Rough Set Theory*, Intelligent Systems Reference Library 168, https://doi.org/10.1007/978-3-030-29566-0_4

4.2 Background

4.2.1 Recommender Systems

Recommender systems use the opinions of a community of users to help individuals in that community more effectively identify content of interest from a potentially overwhelming set of choices [10]. Collaborative filtering is one of the most popular methods of recommender systems and is used in various commercial or research recommender systems. In general, inputs of collaborative filtering is a user-item rating matrix.

In the case of user-based collaborative filtering, similarity score of an active user and other users is measured by some similarity criterion (e.g., Pearson correlation coefficient, cosign similarity, etc.) and the top k users with the highest similarity scores with respect to the active user are selected. The set of the most similar users, called the neighborhood, of the active user is used for predicting rating scores of items that the active user has not rate, and the top t items become the candidate of recommendation to the active user.

4.2.2 Rough Set Theory

Rough set theory, originally proposed by Pawlak [8, 9], provides a mathematical basis of data analysis of nominal datasets. Targets of data analysis by rough set theory is described by decision tables. Here, we use an extended formulation of decision tables by Yao et al. [15] instead of Pawlak's original definition. A decision table DT is defined by the following quintuple:

$$DT = (U, C \cup \{d\}, \{V_a \mid a \in AT\}, \mathbf{R}_{AT}, \rho), \qquad (4.1)$$

where U is a finite and nonempty set of objects, C is a finite and nonempty set of condition attributes, $d \notin C$ is the decision attribute, $\{V_a \mid a \in AT\}$ is a family of sets V_a of attribute values for each attribute $a \in AT = C \cup \{d\}$, and $\rho : U \times AT \to V$ is an information function that assigns a value $\rho(x, a) \in V_a$ of the object $x \in U$ at the attribute $a \in AT$, where $V = \cup_{a \in AT} V_a$. For each attribute $a \in AT$, various relationships defined on the domain V_a are available, e.g., equality relation, order relation, similarity relation, etc., and $\mathbf{R}_{AT} = \{\{R_a\} \mid a \in AT\}$ is a set of families $\{R_a\}$ of binary relations defined on V_a for every attribute $a \in AT$. Note that usual decision tables are implicitly assumed that the family $\{R_a\}$ consists of only the equality relation $=$ defined on V_a [9]. We also assume that every family $\{R_a\}$ includes the equality relation $=$ defined on V_a.

4.2.3 Rough-Set-Based Interrelationship Mining

Rough-set-based interrelationship mining was proposed by Kudo and Murai [6] and it aims at extracting characteristics based on interrelationships between different attributes in a given decision table [7]. To explicitly treat interrelationships between different attributes, the given decision table defined by (4.1) is reformulated as follows:

$$DT_{int} = (U, AT_{int}, V \cup \{0, 1\}, \mathbf{R}_{int}, \rho_{int}), \tag{4.2}$$

where U and V are identical to the original decision table DT. The set $\mathbf{R}_{int}$ of families of binary relations is defined by

$$\mathbf{R}_{AT} \cup \{\{R_{a_i \times b_i}\} \mid \exists a_i, a_i \in C\} \cup \{\{=\} \mid for\ each\ aRb\}, \tag{4.3}$$

where each family $\{R_{a_i \times b_i}\} = \{R^1_{a_i \times b_i}, \ldots, R^{n_i}_{a_i \times b_i}\}$ consists of $n_i (n_i \geq 0)$ binary relations defined on $V_{a_i} \times V_{b_i}$. The expression aRb is defined later.

The set AT_{int} of attributes is defined by

$$AT_{int} = AT \cup \{aRb \mid \exists R \in R_{a \times b}, R(a, b) \neq \emptyset\}, \tag{4.4}$$

and each expression aRb is called an interrelated condition attribute that describes an interrelationship between two condition attribute a and b explicitly. The set $R(a, b)$ is the set of objects that supports the interrelationship between two condition attributes a and b by the binary relation R and called the support set of the interrelationship between a and b. The support set $R(a, b)$ is defined by

$$R(a, b) = \{x \in U \mid (\rho(x, a), \rho(x, b)) \in R\}, \tag{4.5}$$

The set $AT = C \cup \{d\}$ is identical to the original decision table.

Finally, the information function ρ_{int} is defined by

$$\rho_{int}(x, c) = \begin{cases} \rho(x, c), & \text{if } c \in AT, \\ 1, & \text{if } c = aRb \text{ and } x \in R(a, b) \\ 0, & \text{if } c = aRb \text{ and } x \notin R(a, b) \end{cases} \tag{4.6}$$

4.2.4 Yamawaki et al.'s Recommendation Method

In this subsection, we briefly review Yamawaki et al.'s recommendation method [13, 14].

This method is an extension of collaborative filtering and it mainly consists of the following three parts:

(1) Extraction of users' preference patterns as interrelated condition attributes in interrelationship mining,

(2) Evaluation of the similarity of preference patterns between the active user and other users, and

(3) Prediction of active user's rating scores and making a list of recommended items to the active user.

Suppose that $U = \{u_1, \ldots, u_l\}$ is a finite set of users and $P = \{p_1, \ldots, p_m\}$ is a finite set of items. Let R be a user-item rating matrix, where R is an $l \times m$ matrix and the element $r_{i,j} \in Range \cup \{*\}$ at the i-th row and the j-th column of R is the rating score of the j-th item by the i-th user. The symbol $*$ means that the user does not rate the item. Yamawak et al.'s method is as follows [14]:

(Extraction of preference patterns)
1. Construct n pairs of items $(a_1, b_1), \ldots, (a_n, b_n)$ by randomly selecting items from the set of items that the active user $x \in U$ has rated.
2. For each pair (a, b) constructed in step 1, make an interrelated attribute $a \le b$ by using a binary relation $\le$ defined on $V_a \times V_b$.
3. For each user $u \in U$, set the values of the interrelated attributes $a \le b$ as follows:

$$\rho_{int}(u, a \le b) = \begin{cases} 0, & \text{if } \rho(u, a) > \rho(u, b), \\ 1, & \text{if } \rho(u, a) \le \rho(u, b), \\ *, & \text{if } \rho(u, a) = * \text{ or } \rho(u, b) = *. \end{cases} \qquad (4.7)$$

The value of the interrelated attribute $a \le b$ for the active user $x \in U$ represents a preference pattern of the active user. The value of $a \le b$ is 0 means that the user prefers the item a to the item b. The value of $a \le b$ is 1 means that either the user prefers b to a or the user likes (or dislikes) both a and b at the same level.

(Evaluation of similarity of preference patterns)
4. Compare the preference patterns between the active user $x \in U$ and the other users $u \in U$ as follows:

$$sim(x, u) = \frac{| \{a \le b \in AT_{int} \mid \rho(x, a \le b) = \rho(u, a \le b)\} |}{n}, \qquad (4.8)$$

where $|X|$ means the cardinality of the set X. The similarity score $sim(x, u)$ represents the ratio of the user's preference patterns that is consistent with the active user's preference patterns.

(Prediction of active user's rating scores and recommendation)
5. Select top k users with highest similarity scores as the neighborhood N of the active user $x \in U$.
6. For each item $p \in P$ that the active user $x \in U$ does not have the rating score, estimate active user's rating score of the item p as follows:

$$pred(x, p) = \bar{r}_x + \frac{\sum_{u \in N_p} sim(x, u) \times (r_{u,p} - \bar{r}_u)}{\sum_{u \in N_p} sim(x, u)}, \qquad (4.9)$$

where $\bar{r}_x$ and $\bar{r}_u$ are the average scores of the active user x and the user u, respectively, and the set

$$N_p = \{u \in N \mid r_{u,p} \neq *\}$$

is the set of users in the neighborhood N that the user has rated the item p.

7. Select t items with highest estimated rating scores and make a list of recommended items to the active user.

Yamawaki et al. [13] also proposed a revised method of extraction of preference patterns. This method considers not only the order relationship between rating scores but also the difference of rating scores.

4.3 Proposed Method

In this section, we introduce a method to directly set user's preference patterns, and report a prototype of tourist spot recommender system based on the proposed method.

4.3.1 Direct Setting of Preference Patterns

As we reviewed in Sect. 4.2.4 Yamawaki et al.'s method makes users' preference patterns from the users' rating scores of items. However, this extraction method of preference patterns has limitation of representation ability and the following situation is not representable. Suppose that a, b, and c are three items and a user has the following preference patters:

- The user prefers the item b to the item a,
- The user prefers c to b, and
- The user prefers a to c.

We think that this kind of "cyclic" preference patterns also represent user's Kansei evaluation of items. To treat such characteristics of preference patterns, we introduce a method to directly set preference patterns by users. Our method is based on partial pairwise comparison. In the traditional pairwise comparison (e.g., Thurstone's pairwise comparison [11]), all possible pairs of items are required to evaluate by a subject, however, it is actually impossible if there are many items. We then propose the following method to directly set users' preference patterns:

1. Set the number n of constructed preference patterns.
2. $PP = \emptyset$
3. Create a set of pairs of items

$$Q = \{(a_1, b_1), \ldots, (a_n, b_n)\},$$

where there are no pairs that both (a_i, b_i) and (b_i, a_i) are in the set Q.

4. For each pair (a_i, b_i) $(1 \le i \le n)$, the user selects either a_i or b_i intuitively. If a_i is selected, add the preference pattern $a_i > b_i$ to the set of preference patterns PP. Otherwise, add $a_i < b_i$ to PP.

5. Output PP as the set of the user's preference patterns.

4.3.2 Recommendation Method by Directly Setting of Preference Patterns

Our recommendation method is based on Yamawaki et al.'s recommendation method in Sect. 4.2.4. Extraction of preference patterns, steps 1–3 in Sect. 4.2.4, are replaced to the proposed method in Sect. 4.3.1. For evaluation of similarity between the active user $x \in U$ and the other user $u \in U$ we use the following criterion instead of the Eq. (4.8):

$$sim(x, y) = |PP_x \cap PP_u| \tag{4.10}$$

where PP_x and PP_u are the sets of preference patterns of the active user x and the other user u, respectively. Step 5 for selection of the neighborhood is identical to the one in Sect. 4.2.4.

Because our method does not use the rating information, selection of recommended items to the active user, steps 6 and 7, are different from the ones in Sect. 4.2.4. Therefore, these two steps are replaced by the following two steps:

6. For each user $u \in N$ in the neighborhood of the active user, make a set of items P_u that are most frequently selected by the user u.

7. Select t items randomly from the set $\bigcup_{u \in N} P_i$.

4.3.3 A Prototype of Recommender System

We implemented a prototype of recommender system based on the proposed approach. This system recommends tourist spots in Hokkaido prefecture, Japan, from 43 tourist spots that we inputted to the system. For each tourist spot, we selected one photo from two web pages [4, 5] that open various photos of many tourist spots in Hokkaido.

Figure 4.1 shows a screenshot of the implemented system. The user of this system is presented a pair of photos and is asked which tourist spot she/he wants to visit. For example, in Fig. 4.1, the user is asked which tourist spot (left: Sapporo Clock Tower, right: Night view of Hakodate city) the user wants to visit. The user answers intuitively by clicking one of the two photos. By repeating this question-answering 10 times, the user's preference patterns are obtained and at most 5 tourist spots are recommended by comparing the user's preference patterns and other users' preference patterns that have already inputted in the system.

Fig. 4.1 Screenshot of the recommender system

4.4 Experiments

We conducted a pre-experiment and an evaluation experiment to evaluate usefulness of the proposed method. Note that all experiments were organized with respect to Ethics Regulations related to Research with Human Subjects in Muroran Institute of Technology.

4.4.1 Pre-experiment

The purpose of the pre-experiment is to collect other users' sets of preference patterns in the recommender system. 31 subjects attended to the pre-experiment. In the pre-experiment, each subject was presented 50 pairs of photos and answered intuitively which tourist spot in the pair she/he wants to visit. By this pre-experiment, the sets of preference patterns of 50 subjects were obtained and these sets were used as the other users' sets of preference patterns in the evaluation experiment.

4.4.2 Evaluation Experiment

In the evaluation experiment, 10 subjects who did not attend the pre-experiment attended. In this experiment, the number of the subject's preference patterns was $n = 10$, the number of neighborhood users was $k = 3$ as almost 1/10 of the users in the system, and the number of recommended tourist spots was $t = 5$. In the case that the number of candidates of recommended tourist spots is less than five, all candidates were recommended.

For comparison, a recommender system that randomly provides at most 5 tourist spots as the recommendation result was used. To avoid the effect of difference of presented pairs and the order effect, after 10 times question-answering by the subject, recommendation results by the two systems were presented simultaneously without informing which result is by the proposed system. For each result, the subject evaluated the recommended tourist spots by 5-point scale rating score (1: tourist spot that the user does not want to visit, 3: neither, 5: tourist spot that the user wants to visit).

As the result, the average evaluation score of the proposed system was 3.87 and the average score of the random recommender was 3.28. Note that the average similarity score between the subject and other user in the proposed system was 2.57, which means that, on average, the set of preference patterns for each subject has 2 or 3 common patterns with other users in the system.

The score of the proposed system was higher than the score of the random recommender; however, there was no statistically significant difference by paired samples t-test ($p = 0.08 > 0.05$). One reason of this result is that the number of subjects was small and further evaluation experiments with many subjects are needed.

4.5 Conclusions

In this chapter, we introduced a recommendation method that the active user inputs her/his preference patterns directly. The proposed method is an improvement of Yamawaki et al.'s method to extract users' preference patterns and our method is applicable even though the user has a kind of "cyclic" preference patterns. We also implemented a tourist spot recommender system as a prototype of recommender system based on the proposed approach.

For future works, we need to conduct evaluation experiments again with many subjects and users in the system. It is expected that the number of subjects and users will strongly affect the experiment results. In the prototype system, the number of tourist spots was 43 and it is very small as a recommender system. Increasing the number of tourist spots is also important for improving the prototype system.

Moreover, effects of parameter setting (the number of presented pairs, the size of neighborhood, the number of recommended items, etc.), similarity criterion, and selection method of recommended items are also considered for revision.

References

1. Bradley, K., Smyth, B.: Improving recommendation diversity. In: Proceedings of the 12th National Conference on Artificial Intelligence and Cognitive Science (AICS-01), Maynooth, Ireland, pp. 75–84 (2001)
2. Gan, M.X., Jiang, R.: Constructing a user similarity network to remove adverse influence of popular objects for personalized recommendation. Expert Syst. Appl. **40**, 4044–4053 (2013)

3. Gan, M.X., Jiang, R.: Improving accuracy and diversity of personalized recommendation through power law adjustments of user similarities. Decis. Support Syst. **55**, 811–821 (2013)
4. Hokkaido muryo shashin sozai-shu do photo. http://photo.hokkaido-blog.com (in Japanese)
5. Hokkaido prefecture free photos. https://www.photock.org/list/pr/hokkaido
6. Kudo, Y., Murai, T.: A plan of interrelation-ship mining using rough sets. In: Proceedings of the 29th Fuzzy System Symposium, pp. 33–36 (2013) (in Japanese)
7. Kudo, Y., Murai, T.: A review on rough set-based interrelationship mining. In: Torra, V., Dahlbom, A., Narukawa, Y. (eds.) Fuzzy Sets, Rough Sets, Multisets and Clustering, pp. 257–275, Springer, Berlin (2017)
8. Pawlak, Z.: Rough sets. Int. J. Comput. Inf. Sci. **11**, 341–356 (1982)
9. Pawlak, Z.: Rough Sets: Theoretical Aspects of Reasoning about Data. Kluwer Academic Publisher, Dordrecht (1991)
10. Resnick, P., Varian, H.R.: Recommender systems. Commun. ACM **40**, 56–58 (1997)
11. Thurstone, L.L.: A Law of comparative judgment. Psychol. Rev. **34**, 273–286 (1927)
12. Vozalis, M.G., Markos, A.I., Margaritis, K.G.: A hybrid approach for improving prediction coverage of collaborative filtering. In: Proceedings of AIAI 2009, pp. 491–498. Springer (2009)
13. Yamawaki, J., Kudo, Y., Murai, T.: An improved recommendation method based on interelationship mining and collaborative filtering. In: Proceedings of JSKE19, Tokyo, Japan, B26 (2017) (in Japanese)
14. Yamawaki, J., Kudo, Y., Murai, T.: Proposal of a recommendation method based on Interelationship mining and collaborative filtering. In: Proceedings of JSKE12S, Osaka, Japan, 1B-07 (2017) (in Japanese)
15. Yao, Y.Y., Zhou, B., Chen, Y.: Interpreting low and high order rules: a granular computing approach. In: Proceedings of RSEISP 2007, LNCS 4585, pp. 371–380. Springer (2007)
16. Zhang, Z.P., Kudo, Y., Murai, T.: Neighbor selection for user-based collaborative filtering using covering-based rough sets. Ann. Oper. Res. **256**, 359–374 (2017)
17. Zhou, T., Kuscsik, Z., Liu, J.G., Medo, M., Wakeling, J.R., Zhang, Y.C.: Solving the apparent diversity-accuracy dilemma of recommender systems. In: Proceedings of the National Academy of Sciences of the United States of America, vol. 107, pp. 4511–4515 (2010)

Chapter 5
Rough-Set-Based Interrelationship Mining for Incomplete Decision Tables

Abstract This chapter discusses rough-set-based interrelationship mining for incomplete decision tables. Rough-set-based interrelationship mining enables to extract characteristics by comparing the values of the same object between different attributes. To apply this interrelationship mining to incomplete decision tables with null values, in this study, we discuss the treatment of null values in interrelationships between attributes. We introduce three types of null values for interrelated condition attributes and formulate a similarity relation by such attributes with these null values.

Keywords Interrelationship mining · Incomplete decision tables · Rough set · Similarity relation

5.1 Introduction

Rough set theory, originally proposed by Pawlak [12, 13], provides a mathematical basis for logical data analysis, and attribute reduction and decision rule extraction are currently useful tools for data mining [14]. Application of rough set theory to incomplete decision tables was first reported by Kryszkiewicz [2], who used null values to represent a lack of values of objects in some condition attributes and introduced similarity relations between objects rather than equivalence relations, i.e., null values weaken the discernibility of objects by comparing attribute values.

Rough-set-based interrelationship mining, proposed by previous studies [4–7, 9, 10], enables to extract characteristics by interrelationships between attributes. We introduced the concept of interrelationships between attributes by comparing values of the same object between different attributes and formulated indiscernibility between objects by the interrelationship between attributes as the indiscernibility based on whether both objects support (or do not support) the interrelationship. We also introduced new attributes, called interrelated condition attributes, to explicitly represent the interrelationship between two condition attributes.

S. Akama et al., *Topics in Rough Set Theory*, Intelligent Systems Reference Library 168, https://doi.org/10.1007/978-3-030-29566-0_5

To apply rough-set-based interrelationship mining to decision tables with null values, i.e., incomplete decision tables, we need to discuss how to treat null values in interrelationships between attributes. We introduced three types of null values for interrelated condition attributes and formulated the similarity relation by such attributes with these null values.

The reminder of this chapter is organized as follows. Section 5.2 reviews rough set theory applied to incomplete decision tables, and Sect. 5.3 reviews rough-set-based interrelationship mining for complete decision tables. In Sect. 5.4, we introduce similarity relations by interrelationship between attributes in incomplete decision tables, and in Sect. 5.5, we describe interrelated condition attributes in such tables. Section 5.6 presents the conclusions of this study. Note that this paper is a revised and extended version of a previous paper [3].

5.2 Rough Sets for Incomplete Decision Tables

In this section, we review rough sets for incomplete decision tables proposed by Kryszkiewicz [2].

5.2.1 Decision Tables and Similarity Relations

The subjects of rough set data analysis are described by *decision tables*. According to the authors' previous study [5, 7, 9, 10], we use a general expression of decision tables used by Yao et al. [15], which can be expressed as

$$DT = (U, AT, \{V_a \mid a \in AT\}, \mathcal{R}_{AT}, \rho), \tag{5.1}$$

where U is a finite and nonempty set of objects; $AT = C \cup \{d\}$ is a finite and non-empty set of attributes (where C is a set of condition attributes and $d \notin C$ is a decision attribute); V_a is a nonempty set of values for $a \in AT$; $\mathcal{R}_{AT} = \{\{R_a\} \mid a \in AT\}$ is a set of sets $\{R_a\}$ of binary relations defined on each V_a; and ρ is an information function $\rho : U \times AT \rightarrow V$ that presents a value $\rho(x, a) \in V_a$ of each object $x \in U$ at the attribute $a \in AT$ (where $V = \bigcup_{a \in AT} V_a$ is the set of values of all attributes in AT).

The set $\{R_a\}$ of binary relations for each attribute $a \in AT$ can contain various binary relations: similarity, dissimilarity, the ordering relation on V_a and typical information tables are implicitly assumed that the set $\{R_a\}$ consists of only the equality relation $=$ on V_a [15]. We also assume that the equality relation $=$ is included in the set $\{R_a\}$ for each attribute $a \in AT$.

A decision table DT is considered as an incomplete decision table if there is at least one condition attribute $a \in C$ for which V_a contains a null value [2]. We denote a null value by the $*$ symbol and assume that the decision attribute d does not have null value, i.e., $* \notin V_d$. We also assume that, for any attribute $a \in AT$ and any binary relation $R \in \{R_a\}$ except for the equality relation $=$, if $* \in V_a$ holds, then the null value $*$ does not relate to other values $v \in V_a$ by the binary relation R, i.e., neither $*Rv$ nor $vR*$ holds.

A similarity relation $SIM(A)$ between objects that are possibly indiscernible in terms of values of attributes $A \subseteq AT$ is expressed as

$$SIM(A) = \left\{ (x, y) \,\middle|\, \begin{array}{l} \forall a \in A, \ \rho(x, a) = \rho(y, a) \text{ or} \\ \rho(x, a) = * \text{ or } \rho(y, a) = * \end{array} \right\}. \tag{5.2}$$

If a pair (x, y) is in $SIM(A)$, then we say that object x is similar to object y in terms of the attribute values in A. It can be clearly observed that such similarity relation is a tolerance relation, i.e., it is reflexive and symmetric; however, it may not be transitive in general. Note that if all attributes $a \in A$ do not have null value, the similarity relation $SIM(A)$ is identical to the indiscernibility relation $IND(A)$ [2]. Moreover, the following property holds for any similarity relation:

$$SIM(A) = \bigcap_{a \in A} SIM(\{a\}). \tag{5.3}$$

Let $S_A(x) = \{y \in U \mid (x, y) \in SIM(A)\}$ be the set of similar objects with the object x by $A \subseteq AT$. The collection of sets $S_A(x)$ for all $x \in U$ comprises a covering of U, denoted $U/SIM(A)$. Each element in $U/SIM(A)$ is called a tolerance class. In particular, the covering $U/SIM(\{d\})$ generated by the decision attribute d becomes a partition of U, and each element in it is called a decision class.

For any set of objects $X \subseteq U$, the lower approximation $\underline{A}(X)$ and upper approximation $\overline{A}(X)$ of X by $A \subseteq AT$ are defined in a manner similar to Pawlak's rough sets:

$$\underline{A}(X) = \{x \in U \mid S_A(x) \subseteq X\}, \tag{5.4}$$
$$\overline{A}(X) = \{x \in U \mid S_A(x) \cap X \neq \emptyset\}. \tag{5.5}$$

5.2.2 Relative Reducts and Decision Rules

To describe relative reducts for incomplete decision tables, we introduce the generalized decision proposed by Kryszkiewicz [2].

Let ∂_A, $A \subseteq C$, be a function $\partial_A : U \to \mathcal{P}(V_d)$, where $\mathcal{P}(V_d)$ is the power set of V_d, defined as follows:

$$\partial_A(x) = \{v \in V_d \mid y \in S_A(x) \text{ and } \rho(y, d) = v\}. \tag{5.6}$$

The function ∂_A is called a generalized decision by A.

Relative reducts for incomplete decision tables are minimal sets of condition attributes that preserve the generalized decision ∂_C by the set of all condition attributes. Formally, a set $A \subseteq C$ is called a relative reduct of DT if and only if A satisfies the following condition:

$$\partial_A = \partial_C \text{ and } \forall B \subset A, \ \partial_B \neq \partial_C. \tag{5.7}$$

We also introduce the relative reducts of DT for each object in U. A set $A \subseteq C$ is called a relative reduct of DT for object $x \in U$ if and only if A satisfies the following condition:

$$\partial_A(x) = \partial_C(x) \text{ and } \forall B \subset A, \ \partial_B(x) \neq \partial_C(x). \tag{5.8}$$

We denote a decision rule generated from a set $A \subseteq C$ of condition attributes in the following form:

$$\bigwedge_{\mathsf{a} \in A}(\mathsf{a}, v) \to \bigvee(\mathsf{d}, w), \tag{5.9}$$

where $v \in V_{\mathsf{a}}$ and $w \in V_{\mathsf{d}}$. The form (a, v) is called a descriptor, and the forms $\wedge_{\mathsf{a} \in A}(\mathsf{a}, v)$ and $\vee(\mathsf{d}, w)$ are called the condition and decision of the decision rule, respectively.

Let r be a decision rule defined by (5.9) with a set of condition attributes $A \subseteq C$, X be a set of objects of property $\wedge_{\mathsf{a} \in A}(\mathsf{a}, v)$, and Y be a set of objects of property $\vee(\mathsf{d}, w)$. The decision rule r is true in DT if and only if $\overline{A}(X) \subseteq Y$ holds. Moreover, r is optimal in DT if and only if r is true in DT and no other rule r' constructed from the proper subset $B \subset A$ is untrue [2]. Note that, similar to Pawlak's rough sets for complete decision tables, discernibility functions can also be used to compute relative reducts in incomplete decision tables [2].

Example 5.1 Table 5.1 shows an example incomplete decision table DT. The decision table consists of a set of eight users of sample products. Here $U = \{u_1, \ldots, u_8\}$ is the set of users, the set of attributes AT consists of the set of condition attributes $C = \{\mathsf{Gender}, \mathsf{Q.1}, \mathsf{Q.2}, \mathsf{Q.3}\}$ that represents user' gender and the answers to questions 1, 2, and 3 about the sample products, and the decision attribute $\mathsf{Purchase}$ that represents user answers to the question about purchase.

Each attribute $\mathsf{a} \in AT$ has the following range of values V_{a}. Note that each condition attribute in this example has null value $*$ in the following range:

$$V_{\mathsf{Gender}} = \{\mathsf{female}, \ \mathsf{male}, \ *\},$$
$$V_{\mathsf{Q.1}} = \{\mathsf{no}, \ \mathsf{yes}, \ *\},$$
$$V_{\mathsf{Q.2}} = V_{\mathsf{Q.3}} = \{\mathsf{v.b.}, \ \mathsf{bad}, \ \mathsf{normal}, \ \mathsf{good}, \ \mathsf{v.g.}, \ *\},$$
$$V_{\mathsf{Purchase}} = \{\mathsf{no}, \ \mathsf{maybe}, \ \mathsf{yes}\},$$

and each attribute $\mathsf{a} \in AT$ has the following set of binary relations on the set V_{a} of values:

Gender : $\{=\}$, Q.1 : $\{=, \prec_{Q.1}, \preceq_{Q.1}\}$,
Q.2 : $\{=, \prec_{Q.2}, \preceq_{Q.2}\}$, Q.3 : $\{=, \prec_{Q.3}, \preceq_{Q.3}\}$,
Purchase : $\{=, \prec_{Purchase}, \preceq_{Purchase}\}$,

where each relation $\prec_a$ is a preference relation that is defined as follows:

$\prec_{Q.1}$:   no $\prec$ yes,
$\prec_{Q.2}$:   v.b. $\prec$ bad $\prec$ normal $\prec$ good $\prec$ v.g.,
$\prec_{Q.3}$:   v.b. $\prec$ bad $\prec$ normal $\prec$ good $\prec$ v.g.,
$\prec_{Purchase}$: no $\prec$ maybe $\prec$ yes.

The contents of Table 5.1 is described by the information function ρ; e.g., $\rho(u1, \text{Gender}) = \text{female}$, and $\rho(u5, \text{Q.1}) = *$.

Here, we consider the similarity relation and relative reduct of Table 5.1. The generalized decision ∂_C of Table 5.1 by the set of all condition attributes $C \subseteq AT$ is defined as follows:

$$\partial_C(u1) = \partial_C(u2) = \partial_C(u3) = \partial_C(u4)$$
$$= \{\text{yes, maybe}\},$$
$$\partial_C(u5) = \partial_C(u6)$$
$$= \{\text{yes, maybe, no}\},$$
$$\partial_C(u7) = \partial_C(u8)$$
$$= \{\text{maybe, no}\}.$$

Let $A = \{\text{Q.1}, \text{Q.2}, \text{Q.3}\}$ be a subset of condition attributes. For each object $u \in U$, the set of similar objects $S_A(u)$ with respect to the similarity relation $SIM(A)$ by (5.2) is as follows:

$$S_A(u1) = \{u1, u5\}, \quad S_A(u2) = \{u2, u5, u6\},$$
$$S_A(u3) = \{u3, u5, u6\}, \quad S_A(u4) = \{u4, u5\},$$
$$S_A(u5) = U, \quad S_A(u6) = \{u2, u3, u5, u6, u7\},$$
$$S_A(u7) = \{u5, u6, u7\}, \quad S_A(u8) = \{u5, u8\}.$$

For example, $u5, u6 \in S_A(u3)$ means that the object $u3$ is indiscernible from the objects $u5$ and $u6$ by the values of attributes in A, which confirms that the generalized decision $\partial_A(u3)$ is {yes, maybe}. Consequently, it is clearly observed that $A = \{\text{Q.1}, \text{Q.2}, \text{Q.3}\}$ is a relative reduct of Table 5.1, i.e., $\partial_A = \partial_C$ and $\forall B \subset A$, $\partial_B \neq \partial_C$ holds.

Similarly, we can also consider a relative reduct of some objects and decision rules. Let $X = \{\text{Q.3}\}$. It can be clearly observed that $S_X(u3) = \{u3, u5, u6\}$ and it provides $\partial_X(u3) = \{\text{yes, maybe}\} = \partial_C(u3)$, i.e., X is a relative reduct of $u3$ in 5.1. By using the relative reduct X and the generalized decision of $u3$, we obtain the following optimal decision rule:

Table 5.1 Incomplete decision table

U	Gender	Q.1	Q.2	Q.3	Purchase
$u1$	Female	Yes	Good	v. g.	Yes
$u2$	Male	No	Good	v. g.	Yes
$u3$	Male	No	*	Good	Yes
$u4$	Female	Yes	Normal	Normal	Yes
$u5$	Female	*	*	*	Maybe
$u6$	Male	No	Good	*	Maybe
$u7$	Male	No	*	Bad	No
$u8$	*	Yes	Good	Normal	No

$$(\text{Q.3, good}) \rightarrow (\text{Purchase, yes}) \vee (\text{Purchase, maybe}).$$

5.3 Interrelationship Mining for Complete Decision Tables

In this section, we review rough-set-based interrelationship mining for complete decision tables [4–7, 9, 10].

5.3.1 Observations and Motivations

Comparison of attribute values between objects is a common basis of rough set data analysis. For example, in Pawlak's rough set, the equality of attribute values for each attribute $a \in A$ is essential for an indiscernibility relation $IND(A)$ by the subset of attributes $A \subseteq AT$. In the dominance-based rough set approach [1], the total preorder relation among attribute values in each attribute provides a dominance relation between objects.

Generally, the domain of such comparison between attribute values is separated into each set of attribute values $V_a, a \in AT$. However, the following characteristics are difficult to describe without comparing the values between different attributes:

- the evaluation score of movie A is better than that of movie B,
- the color of sample A is similar to that of sample B,
- users prefer the design of car A to that of car B.

To treat the above characteristics in the framework of rough sets, we need to extend the domain of binary relations R for comparison of attribute values from each range V_a, i.e., $R \subseteq V_a \times V_a$ to the Cartesian product of two ranges V_a and V_b, i.e., $R \subseteq V_a \times V_b$. This extension enables us to describe the interrelationships between attributes by comparing the attribute values of different attributes in the framework of rough set theory [6].

5.3.2 Interrelationships Between Attributes and Indiscernibiilty of Objects by Interrelationships

Various relations between attribute values can be considered to compare values, e.g., equality, equivalence, order relations, and similarity. These situations between attributes in a given decision table can be formulated as interrelationships as follows.

Let DT be a decision table by (5.1), $a, b \in C$ be two condition attributes with ranges V_a and V_b, respectively, and $R \subseteq V_a \times V_b$ be a binary relation from V_a to V_b. The attributes a and b are interrelated by R if and only if there exists an object $x \in U$ such that $(\rho(x, a), \rho(x, b)) \in R$ holds [6]

We denote the set of objects wherein the values of attributes a and b satisfy the relation R as follows:

$$R(a, b) = \{x \in U \mid (\rho(x, a), \rho(x, b)) \in R\}, \qquad (5.10)$$

and we call the set $R(a, b)$ the support set of the interrelationship between a and b by R.

An interrelationship between two attributes by a binary relation provides a way to compare attribute values between different attributes; however, for simplicity, we also allow the interrelationship between the same attributes.

Indiscernibility relations in a given decision table by interrelationships between attributes are introduced as follows [6]. Let $a, b \in C$ be two condition attributes in a decision table DT. We assume that $a, b \in C$ are interrelated by a binary relation $R \subseteq V_a \times V_b$, i.e., $R(a, b) \neq \emptyset$ holds. The indiscernibility relation on U based on the interrelationship between a and b by R is defined by

$$IND(aRb) = \left\{ (x, y) \;\middle|\; \begin{array}{l} x \in U, y \in U, \text{ and} \\ x \in R(a, b) \text{ iff } y \in R(a, b) \end{array} \right\}. \qquad (5.11)$$

For any objects x and y, $(x, y) \in IND(aRb)$ means that x is not discernible from y relative to whether the interrelationship between the attributes a and b by relation R holds. Any binary relation $IND(aRb)$ on U defined by (5.11) is an equivalence relation, and we can construct equivalence classes from an indiscernibility relation $IND(aRb)$ [6].

5.3.3 Decision Tables for Interrelationship Mining

The decision table DT by (5.1) is reconstructed to treat interrelationships between attributes explicitly by introducing the given binary relations to compare the values of different attributes.

Definition 5.1 ([7, 10]) Let DT be a decision table by (5.1). The decision table DT_{int} for interrelationship mining with respect to DT is defined as follows:

$$DT_{int} = (U, AT_{int}, V \cup \{0, 1\}, \mathcal{R}_{int}, \rho_{int}), \tag{5.12}$$

where U and $V = \bigcup_{a \in AT} V_a$ are identical to DT.

The set AT_{int} is defined by

$$AT_{int} = AT \cup \{a\mathsf{R}b \mid \exists R \in \{R_{a \times b}\}, R(\mathsf{a}, \mathsf{b}) \neq \emptyset\}, \tag{5.13}$$

where $AT = C \cup \{d\}$ is identical to DT and each expression $a\mathsf{R}b$ is a newly introduced condition attribute called an interrelated condition attribute. The set $\{R_{a \times b}\}$ of binary relation(s) is defined below.

The set $\mathcal{R}_{int}$ of sets of binary relations is defined as

$$\mathcal{R}_{int} = \mathcal{R}_{AT} \cup \left\{ \{R_{a_i \times b_i}\} \,\middle|\, \begin{array}{l} R_{a_i \times b_i} \subseteq V_{a_i} \times V_{b_i}, \\ \exists a_i, b_i \in C \end{array} \right\} \tag{5.14}$$
$$\cup \{\{=\} \mid \text{For each } a\mathsf{R}b\},$$

where each set $\{R_{a_i \times b_i}\} = \{R^1_{a_i \times b_i}, \dots, R^{n_i}_{a_i \times b_i}\}$ consists of n_i ($n_i \geq 0$) binary relation(s) defined on $V_{a_i} \times V_{b_i}$.

The information function ρ_{int} is defined by

$$\rho_{int}(x, \mathsf{c}) = \begin{cases} \rho(x, \mathsf{c}), & \text{if } \mathsf{c} \in AT, \\ 1, & \mathsf{c} = a\mathsf{R}b \text{ and } x \in R(\mathsf{a}, \mathsf{b}), \\ 0, & \mathsf{c} = a\mathsf{R}b \text{ and } x \notin R(\mathsf{a}, \mathsf{b}). \end{cases} \tag{5.15}$$

Each interrelated condition attribute $a\mathsf{R}b$ represents whether each object $x \in U$ supports the interrelationship between the attributes $\mathsf{a}, \mathsf{b} \in C$ by the binary relation $R \subseteq V_a \times V_b$ [6]. Therefore, interrelated condition attributes are nominal attributes and we only treat the equality relation to compare attribute values of different objects for each interrelated condition attribute.

Note that the Eq. (5.14) means that binary relations for comparing attribute values between different attributes are assumed to be given. In general, comparison of attribute values between different attributes is needed to treat carefully and we need to determine whether two attributes in a given decision table are comparable in some sense based on some background knowledge.

However, comparability between attribute values depends on the "meaning of dataset", in other words, semantics of interrelationship mining and the formulation of semantics is one of the most important issues for rough-set-based interrelationship mining [10].

The following property guarantees the equivalence of representation ability between the interrelationship between attributes a and b by R and the corresponding interrelated condition attribute $a\mathsf{R}b$.

Proposition 5.1 ([6]) *Let DT be a decision table by (5.1) and DT_{int} be a decision table by (5.12). The following equality holds.*

$$IND_{DT}(\mathsf{a}R\mathsf{b}) = IND_{DT_{int}}(\{\mathsf{a}R\mathsf{b}\}), \tag{5.16}$$

where $IND_{DT}(\mathsf{a}R\mathsf{b})$ is the indiscernibility relation in DT with respect to the inter-relationship between a and b by R defined by (5.11), and $IND_{DT_{int}}(\{\mathsf{a}R\mathsf{b}\})$ is the indiscernibility relation in DT_{int} constructed from a singleton $\{\mathsf{a}R\mathsf{b}\}$.

The authors have discussed effectiveness of interrelated condition attributes in relative reducts of complete decision tables [8]. The following property guarantees that interrelated condition attributes in relative reducts enhance representation ability of decision rules generated from relative reducts.

Proposition 5.2 ([8]) *Suppose $A \subseteq AT_{int}$ is a relative reduct of the decision table DT_{int} for interrelationship mining. For any two condition attributes $\mathsf{a}, \mathsf{b} \in A$, let a set A' be either $A' = (A \setminus \{\mathsf{a}\}) \cup \{\mathsf{a}R\mathsf{b}\}$ or $A' = (A \setminus \{\mathsf{b}\}) \cup \{\mathsf{a}R\mathsf{b}\}$ and A' is also a relative reduct of DT_{int}. The following inequality holds:*

$$ACov(A) \leq ACov(A'). \tag{5.17}$$

Note that $ACov(\cdot)$ is an evaluation criterion for relative reducts proposed by the authors [11] based on the roughness of partitions generated from the relative reducts. The criterion $ACov(\cdot)$ is defined by

$$ACov(A) = \frac{|\mathcal{D}|}{\displaystyle\sum_{[x]_A \in U/IND(A)} |\{D_i \in \mathcal{D} \mid [x]_A \cap D_i \neq \emptyset\}|}, \tag{5.18}$$

where $|X|$ is the cardinality of the set X, and $\mathcal{D} = U/IND(\{\mathsf{d}\})$ is the set of decision classes of the given decision table. For every relative reduct $A \subseteq AT_{int}$, the score $ACov(A)$ represents the average of coverage scores of decision rules generated from A, the decision attribute d, and every object $x \in U$ [11].

5.4 Interrelationships Between Attributes in Incomplete Decision Tables

In this section, we revise the concept of interrelationships between condition attributes to treat null value in the interrelationships and introduce similarity relations based on the interrelationship between attributes.

5.4.1 Three Cases in Which Interrelationships are Not Available by Null Value

Here, let $\mathsf{a}, \mathsf{b} \in C$ be two condition attributes and R be a binary relation defined on $V_\mathsf{a} \times V_\mathsf{b}$. Similar to the case of binary relations R_a on V_a, we do not treat the relation between null value $*$ and other values by R, which causes the following three cases in which we cannot consider the interrelationship between a and b by R:

1. The vaule of a is null but that of b is not null: $\rho(x, a) = *$ and $\rho(x, b) \neq *$,
2. The vaule of b is null but that of a is not null: $\rho(x, a) \neq *$ and $\rho(x, b) = *$,
3. The values of both a and b are null: $\rho(x, a) = \rho(x, b) = *$.

Because the occurrences of null value are different among these three cases, we can consider the three cases separately to treat null value in the interrelationship between a and b by R. Moreover, this fact indicates that each object $x \in U$ fits one of the following three situations:

(a) x supports the interrelationship between a and b by R, i.e., $x \in R(a, b)$,
(b) x does not support the interrelationship between a and b by R,
(c) The values of x at a and b are not comparable by one of the above three cases.

The situation (c) causes difficulty of interrelationship between attributes, and therefore, in the next section, we discuss an approach to treat the situation (c) and introduce similarity relations by interrelationship between attributes.

5.4.2 Similarity Relation by Interrelationship Between Attributes

According to the discussion in Sect. 5.4.1, we introduce the following three sets of objects:

$$LN(a, b) = \{x \in U \mid \rho(x, a) = * \text{ and } \rho(x, b) \neq *\}, \qquad (5.19)$$
$$RN(a, b) = \{x \in U \mid \rho(x, a) \neq * \text{ and } \rho(x, b) = *\}, \qquad (5.20)$$
$$BN(a, b) = \{x \in U \mid \rho(x, a) = \rho(x, b) = *\}. \qquad (5.21)$$

The sets $LN(a, b)$, $RN(a, b)$, and $BN(a, b)$ correspond to the cases 1, 2, and 3, respectively. We denote the union of $LN(a, b)$, $RN(a, b)$, and $BN(a, b)$ as $NULL(a, b) \overset{\text{def}}{=} LN(a, b) \cup RN(a, b) \cup BN(a, b)$. The set $NULL(a, b)$ corresponds to the set of of objects in the situation (c).

For every binary relation R defined on $V_a \times V_b$, the domain of R can be naturally extended to the case that V_a and V_b include the null value $*$. As we mentioned in Sect. 5.4.1, for any two condition attributes a, b $\in C$ and any binary relation R defined on $V_a \times V_b$, we do not treat the relationship between the null value $*$ and all values $v \in V_a$ and $w \in V_b$ by R, i.e., neither $vR*$ nor $*Rw$ holds. Note that this modification does not affect the definition of support set $R(a, b)$ by (5.10).

We then represent the set of objects that does not support the interrelationship between a and b by R as follows:

$$R^c(\mathsf{a}, \mathsf{b}) = U - (R(\mathsf{a}, \mathsf{b}) \cup NULL(a, b)). \tag{5.22}$$

Note that the set $R^c(\mathsf{a}, \mathsf{b})$ defined by (5.22) corresponds to the support set of the interrelationship between a and b by the binary relation $R^c \stackrel{\text{def}}{=} (U_\mathsf{a} \times U_\mathsf{b}) - R$, where the set $U_\mathsf{i} \stackrel{\text{def}}{=} \{x \in U \mid \rho(x, \mathsf{i}) \neq *\}$ is a set objects in which the value of the attribute i is not null.

The following is obvious from the definitions of the support set $R(\mathsf{a}, \mathsf{b})$, nonsupport set $R^c(\mathsf{a}, \mathsf{b})$, and $LN(\mathsf{a}, \mathsf{b})$, $RN(\mathsf{a}, \mathsf{b})$, and $BN(\mathsf{a}, \mathsf{b})$.

Lemma 5.1 *The collection S of sets defined as*

$$S = \left\{ \begin{array}{l} R(\mathsf{a}, \mathsf{b}), R^c(\mathsf{a}, \mathsf{b}), LN(\mathsf{a}, \mathsf{b}), \\ RN(\mathsf{a}, \mathsf{b}), BN(\mathsf{a}, \mathsf{b}) \end{array} \right\} - \{\emptyset\} \tag{5.23}$$

comprises a partition of U.

The proof of Lemma 5.1 can be found in Appendix.

Here, we introduce a similarity relation by the interrelationship between attributes.

Definition 5.2 Let DT be a given decision table, and S be a family of object sets by (5.23). Suppose that two condition attributes $\mathsf{a}, \mathsf{b} \in C$ are interrelated by the binary relation $R \subseteq V_\mathsf{a} \times V_\mathsf{b}$, i.e., $R(\mathsf{a}, \mathsf{b}) \neq \emptyset$ holds.

The similarity relation on U by interrelationship between a and b by R is defined as follows:

$$SIM(\mathsf{a}R\mathsf{b})$$
$$= \left\{ (x, y) \middle| \begin{array}{l} \exists S \in S, \ x, y \in S, \ \text{or} \\ x \in NULL(\mathsf{a}, \mathsf{b}) \ \text{and} \\ y \notin NULL(\mathsf{a}, \mathsf{b}), \ \text{or} \\ x \notin NULL(\mathsf{a}, \mathsf{b}) \ \text{and} \\ y \in NULL(\mathsf{a}, \mathsf{b}) \end{array} \right\}. \tag{5.24}$$

It is easy to confirm that the binary relation $SIM(\mathsf{a}R\mathsf{b})$ by the interrelationship between a and b by R is a tolerance relation on U.

Theorem 5.1 *Let $\mathsf{a}, \mathsf{b} \in C$ be two condition attributes in a given decision table DT and $R \subseteq V_\mathsf{a} \times V_\mathsf{b}$ be a binary relation. The binary relation $SIM(\mathsf{a}R\mathsf{b})$ defined by (5.24) satisfies reflexivity and symmetry.*

The proof of Theorem 5.1 can be found in Appendix.

If both attribute a and b do not have null value, the similarity relation $SIM(\mathsf{a}R\mathsf{b})$ is identical to the indiscernibility relation $IND(\mathsf{a}R\mathsf{b})$ by the interrelationship between a and b by a binary relation R, i.e., $IND(\mathsf{a}R\mathsf{b})$ is a special case of $SIM(\mathsf{a}R\mathsf{b})$.

Corollary 5.1 *If both $* \notin V_\mathsf{a}$ and $* \notin V_\mathsf{b}$ hold, then $SIM(\mathsf{a}R\mathsf{b}) = IND(\mathsf{a}R\mathsf{b})$ holds, where $IND(\mathsf{a}R\mathsf{b})$ is the indiscernibility relation by the interrelationship between the attributes defined by (5.11).*

Example 5.2 We introduce two interrelationships between two condition attributes Q.2 and Q.3 in Table 5.1 by comparing the values of these attributes by the following two binary relations defined on $V_{Q.2} \times V_{Q.3}$:

- $\prec_{Q.2 \times Q.3}$: the answer to Q.3 is better than the answer to Q.2,
- $\preceq_{Q.2 \times Q.3}$: The answer to Q.3 is equal to or better than the answer to Q.2.

The range of these two attributes is identical; therefore, we can consider the preference relation, $\prec_{Q.2}$, as the binary relation $\prec_{Q.2 \times Q.3}$ defined on $V_{Q.2} \times V_{Q.3}$ with the following order:

$$\prec_{Q.2 \times Q.3}: \text{ v.b. } \prec \text{ bad } \prec \text{ normal } \prec \text{ good } \prec \text{ v.g.}.$$

Here, we concentrate on the similarity relation between Q.2 and Q.3 by the binary relation $\prec_{Q.2 \times Q.3}$. Hereafter, we omit the indices of binary relations. We construct the support set $\prec$ (Q.2, Q.3) of the interrelationship defined by (5.10) and the other four sets defined by (5.19)–(5.22) as follows:

$$\prec (Q.2, Q.3) = \{u1, u2\},$$
$$\prec^c (Q.2, Q.3) = \{u4, u8\},$$
$$LN(Q.2, Q.3) = \{u3, u7\},$$
$$RN(Q.2, Q.3) = \{u6\},$$
$$BN(Q.2, Q.3) = \{u5\}.$$

We then obtain the set $NULL(Q.2, Q.3) = \{u3, u5, u6, u7\}$ and a partition S by (5.23) as follows:

$$S = \{\{u1, u2\}, \{u4, u8\}, \{u3, u7\}, \{u6\}, \{u5\}\}.$$

The support set $\prec$ (Q.2, Q.3) is not empty, therefore, we can construct the similarity relation $SIM(Q.2 \prec Q.3)$ based on the interrelationship between Q.2 and Q.3 by the binary relation $\prec$ by (5.24). From the constructed similarity relation, we obtain the following results:

- $(u1, u2), (u4, u8) \in SIM(Q.2 \prec Q.3)$: $u1$ and $u2$, and $u4$ and $u8$ are indiscernible from each other, respectively; both $u1$ and $u2$ ($u4$ and $u8$) support (do not support) the interrelationship.
- $(u3, u4), (u4, u5) \in SIM(Q.2 \prec Q.3)$: $u3$ and $u4$, and $u4$ and $u5$ are indiscernible from each other, respectively; an object in the pair adopts one of the three cases that cannot treat the interrelationship between attributes and another does not adopt any of the three cases.
- $(u3, u7) \in SIM(Q.2 \prec Q.3)$: both $u3$ and $u7$ adopt the same case that cannot treat the interrelationship between attributes.
- $(u3, u5) \notin SIM(Q.2 \prec Q.3)$: $u3$ and $u5$ adopt different cases that cannot treat the interrelationship between attributes.

The results $(u3, u4)$, $(u4, u5) \in SIM(Q.2 \prec Q.3)$ but $(u3, u5) \notin SIM(Q.2 \prec Q.3)$ demonstrate that the similarity relation by the interrelationship between attributes generally does not satisfy transitivity.

5.5   Interrelated Attributes for Incomplete Decision Tables

To describe the three cases discussed in Sect. 5.4.1, we introduce three null values $*_l$, $*_r$, and $*_b$ for interrelated condition attributes. These null values $*_l$, $*_r$, and $*_b$ correspond to cases 1, 2, and 3, respectively.

Definition 5.3 Let DT be a decision table by (5.1), and DT_{int} be a decision table for interrelationship mining with respect to the DT defined by (5.12). We redefine the information function ρ_{int} for interrelationship mining defined by (5.15). The redefined information function ρ_{int} is expressed as follows:

$$\rho_{int} : U \times AT_{int} \rightarrow V \cup \{0, 1\} \cup \{*_l, *_r, *_b\} \tag{5.25}$$

such that
$$
\rho_{int}(x, \mathsf{c})
= \begin{cases}
\rho(x, \mathsf{c}), & \text{if } \mathsf{c} \in AT, \\
1, & \mathsf{c} = \mathsf{aRb} \text{ and } x \in R(\mathsf{a}, \mathsf{b}), \\
0, & \mathsf{c} = \mathsf{aRb} \text{ and } x \in R^c(\mathsf{a}, \mathsf{b}), \\
*_l, & \mathsf{c} = \mathsf{aRb} \text{ and } x \in LN(\mathsf{a}, \mathsf{b}), \\
*_r, & \mathsf{c} = \mathsf{aRb} \text{ and } x \in RN(\mathsf{a}, \mathsf{b}), \\
*_b, & \mathsf{c} = \mathsf{aRb} \text{ and } x \in BN(\mathsf{a}, \mathsf{b}).
\end{cases}
\tag{5.26}
$$

Note that there exists various interpretations of the null value $*$, e.g., missing value, nondeterministic value, and do not care. We interpret the newly introduced null value $*_l$ ($*_r$, $*_b$) as "the value of the interrelated condition attribute is not determined by the null value of condition attribute(s) at left (right, both) side(s)." This interpretation does not depend on the interpretation of null value $*$ that causes the occurrence of null values $*_l$, $*_r$, and $*_b$.

Similar to the interrelationship mining for complete decision tables, we can represent the similarity relation based on an interrelationship between attributes using the corresponding interrelated condition attribute.

Definition 5.4 Let DT_{int} be a decision table for the interrelationship mining defined by (5.12) and assume that DT_{int} has a revised information function ρ_{int} defined by (5.26). For any interrelated condition attribute aRb, a similarity relation based on the singleton $\{\mathsf{aRb}\}$ is defined as follows:

$$
\begin{aligned}
&SIM(\{\mathsf{aRb}\}) \\
&= \left\{ (x, y) \left| \begin{array}{l} \rho_{int}(x, \mathsf{aRb}) = \rho_{int}(y, \mathsf{aRb}), \\ \text{or} \\ \rho_{int}(x, \mathsf{aRb}) \in \{1, 0\} \text{ and} \\ \rho_{int}(y, \mathsf{aRb}) \in \{*_l, *_r, *_b\}, \\ \text{or} \\ \rho_{int}(x, \mathsf{aRb}) \in \{*_l, *_r, *_b\} \\ \text{and } \rho_{int}(y, \mathsf{aRb}) \in \{1, 0\}. \end{array} \right. \right\}.
\end{aligned}
\tag{5.27}
$$

It is easy to confirm that the binary relation $SIM(\{\mathsf{aRb}\})$ from an interrelated condition attribute aRb by (5.27) is a tolerance relation.

Theorem 5.2 *Let DT_{int} be a decision table for the interrelationship mining defined by (5.12) and assume that DT_{int} has a revised information function ρ_{int} defined by (5.26). For any interrelated condition attribute $\mathsf{aRb} \in AT_{int}$, the binary relation $SIM(\{\mathsf{aRb}\})$ defined by (5.27) satisfies reflexivity and symmetry.*

The proof of Theorem 5.2 can be found in Appendix.
The newly introduced null values $*_l$, $*_r$, and $*_b$ are mutually discernible.

Corollary 5.2 *Let $SIM(\{\mathsf{aRb}\})$ be the similarity relation defined by (5.27), and $i, j \in \{l, r, b\}$. For any object $x, y \in U$, if $\rho(x, \mathsf{aRb}) = *_i$, $\rho(y, \mathsf{aRb}) = *_j$, and $i \neq j$ hold, then $(x, y) \notin SIM(\{\mathsf{aRb}\})$ holds.*

Similar to the indiscernibility relations in complete decision tables by Proposition 5.1, the similarity relation by the interrelationship between attributes is perfectly representable by the similarity relation with respect to the corresponding interrelated condition attribute.

Theorem 5.3 *Let DT be a decision table by (5.1) and DT_{int} be a decision table for interrelationship mining by (5.12) with the redefined information function by (5.26). The following equality holds.*

$$
SIM_{DT}(\mathsf{aRb}) = SIM_{DT_{int}}(\{\mathsf{aRb}\}),
\tag{5.28}
$$

where $SIM_{DT}(\mathsf{aRb})$ is the similarity relation in DT with respect to the interrelationship between a and b by R defined by (5.24), and $SIM_{DT_{int}}(\{\mathsf{aRb}\})$ is the similarity relation in DT_{int} constructed from the singleton $\{\mathsf{aRb}\}$ of the interrelated condition attribute aRb by (5.27).

The proof of Theorem 5.3 can be found in Appendix.
Similar to the case of similarity relation by the interrelationship between attributes by Corollary 5.1, the similarity relation $SIM(\{\mathsf{aRb}\})$ by the interrelated condition attribute aRb can be considered as the indiscernibility relation $IND(\{\mathsf{aRb}\})$, if attributes a and b do not have null value.

Corollary 5.3 *If both $* \notin V_\mathsf{a}$ and $* \notin V_\mathsf{b}$ hold, then $SIM_{DT_{int}}(\{\mathsf{aRb}\}) = IND_{DT_{int}}(\{\mathsf{aRb}\})$ holds.*

By Theorem 5.3 and the tolerance of similarity relation by (5.3), we can combine the similarity relation $SIM(\{aRb\})$ by the interrelated condition attribute aRb with the similarity relation $SIM(A)$, $A \subseteq C$, defined by (5.2). Moreover, by Corollary 5.3 and the property of similarity relation [2], the combined similarity relation can be considered an indiscernibility relation if there is no null value in the ranges of the attributes in $A \cup \{aRb\}$ used to construct the similarity relation.

Therefore, the results of this study enable us to flexibly treat the similarity between objects with respect to both characteristics by comparing the values of the same attributes and by the interrelationship between attributes.

Example 5.3 Here, we introduce two interrelated condition attributes Q.2$\prec$Q.3 and Q.2$\preceq$Q.3 based on binary relations $\prec$ and $\preceq$ defined on $V_{Q.2} \times V_{Q.3}$, respectively. We can use the results of Example 5.2 to construct the interrelated condition attribute Q.2$\prec$Q.3 and define the value for each object. For the interrelated condition attribute Q.2$\preceq$Q.3, the support set $\preceq$ (Q.2, Q.3) defined by (5.10) and the non-support set $\preceq^c$ (Q.2, Q.3) defined by (5.22) are as follows, respectively:

$$\preceq (Q.2, Q.3) = \{u1, u2, u4\},$$
$$\preceq^c (Q.2, Q.3) = \{u8\}.$$

The sets $LN(Q.2, Q.3)$, $RN(Q.2, Q.3)$, and $BN(Q.2, Q.3)$ are identical to the case of relation $\prec_{Q.2 \times Q.3}$ in Example 5.2.

We then construct the two interrelated condition attributes Q.2$\prec$Q.3 and Q.2$\preceq$Q.3 using the five sets for each binary relation $\prec$ and $\preceq$, respectively. Table 5.2 is the incomplete decision table with the two interrelated condition attributes.

Similar to Example 5.1, we consider the similarity relation and relative reduct of Table 5.2. Let $C_{int} = C \cup \{Q.2 \prec Q.3, Q.2 \preceq Q.3\}$. The generalized decision $\partial_{C_{int}}$ of Table 5.1 by the set of all condition attributes $C \subseteq AT$ is defined as follows:

$$\partial_{C_{int}}(u1) = \partial_{C_{int}}(u2) = \partial_{C_{int}}(u4) = \partial_{C_{int}}(u6)$$
$$= \{yes, maybe\},$$
$$\partial_{C_{int}}(u3) = \{yes\},$$
$$\partial_{C_{int}}(u5) = \{yes, maybe, no\},$$
$$\partial_{C_{int}}(u7) = \{no\},$$
$$\partial_{C_{int}}(u8) = \{maybe, no\}.$$

In Table 5.2, the object $u3$ is discernible from $u6$ by the difference in the null values in the interrelated condition attributes, i.e., $*_l$ at $u3$ and $*_r$ at $u6$. Similarly, $u7$ is discernible from $u6$ and these differences provide the above revision of the generalized decision, i.e., $\partial_{C_{int}}(u3) = \{yes\}$ and $\partial_{C_{int}}(u7) = \{no\}$, respectively.

Table 5.2 Incomplete decision table for interrelationship mining

U	Gender	Q.1	Q.2	Q.3	Q.2≺Q.3	Q.2≼Q.3	Purchase
$u1$	Female	Yes	Good	v. g.	1	1	Yes
$u2$	Male	No	Good	v. g.	1	1	Yes
$u3$	Male	No	*	Good	$*_l$	$*_l$	Yes
$u4$	Female	Yes	Normal	Normal	0	1	Yes
$u5$	Female	*	*	*	$*_b$	$*_b$	Maybe
$u6$	Male	No	Good	*	$*_r$	$*_r$	Maybe
$u7$	Male	No	*	Bad	$*_l$	$*_l$	No
$u8$	*	Yes	Good	Normal	0	0	No

We construct the similarity relation $SIM(\{Q.2 \prec Q.3\})$ with respect to the singleton of the interrelated condition attribute by (5.27). By Theorem 5.3, the constructed similarity relation $SIM(\{Q.2 \prec Q.3\})$ correctly reflects all relationships between objects by the similarity relation $SIM_{DT}(Q.2 \prec Q.3)$ by interrelationship between Q.2 and Q.3 with the binary relation $\prec$ in Example 5.2. For example, similar to the above discussion, the null values of $u3$ and $u5$ are different in Table 5.2, i.e., $*_l$ at $u3$ and $*_b$ at $u5$, which provides $(u3, u4), (u4, u5) \in SIM(\{Q.2 \prec Q.3\})$ but $(u3, u5) \notin SIM(\{Q.2 \prec Q.3\})$.

Here, we consider the relative reduct of Table 5.2. Let $A = \{Q.1, Q.3, Q.2 \preceq Q.3\}$. For each object $u \in U$, the set of similar objects $S_A(u)$ with respect to the similarity relation $Sim(A)$ by (5.2) is as follows, respectively:

$$S_A(u1) = \{u1, u5\}, \qquad S_A(u2) = \{u2, u5, u6\},$$
$$S_A(u3) = \{u3\}, \qquad\qquad S_A(u4) = \{u4, u5\},$$
$$S_A(u5) = \begin{Bmatrix} u1, u2, u4, \\ u5, u8 \end{Bmatrix}, \quad S_A(u6) = \{u2, u6\},$$
$$S_A(u7) = \{u7\}, \qquad\qquad S_A(u8) = \{u5, u8\}.$$

If we remove some attribute from A, e.g., Q.3, the subset $A' = \{Q.1, Q.2 \preceq Q.3\}$ cannot keep the generalized decision by C_{int}, e.g., $S_{A'}(u7) = \{u2, u3, u7\}$ and $\partial_{A'}(u7) = \{yes, no\} \neq \partial_{C_{int}}(u7)$. Consequently, it is easy to observe that $A = \{Q.1, Q.3, Q.2 \preceq Q.3\}$ is a relative reduct of Table 5.2.

Similarly, we can obtain relative reducts of an object, which are used to express optimal decision rules. Let $X = \{Q.3, Q.2 \preceq Q.3\}$. It can be clearly observed that $S_X(u7) = \{u7\}$ and $\partial_X(u7) = \{no\} = \partial_{C_{int}}(u7)$, and any subsets cannot keep the generalized decision, e.g., if $X' = \{Q.3\}$, then $S_{X'}(u7) = \{u5, u6, u7\}$ and $\partial_{X'}(u7) = \{maybe, no\} \neq \partial_{C_{int}}(u7)$. Therefore, X is a relative reduct of $u7$. We then obtain an optimal decision rule of $u7$ as follows:

$$(Q.3, bad) \wedge (Q.2 \preceq Q.3, *_l) \rightarrow (Purchase, no).$$

5.6 Conclusions

In this study, we applied rough-set-based interrelationship mining to incomplete decision tables. First, we observed three cases in which the interrelationship between attributes unavailable by null values and introduced similarity relations based on the interrelationship between attributes. Next, we introduced three types of null values that correspond to the above three cases and demonstrated that the similarity relation by the interrelationship between attributes is perfectly representable by the similarity relation with respect to the corresponding interrelated condition attribute.

In future, we plan to refine the theoretical aspects of the proposed approach and apply it to real-life data analysis. In particular, how to determine the existence or non-existence of interrelationships between condition attributes and how to construct interrelated condition attributes from the scratch by generating binary relations between attribute values are very important issues for applying rough-set-based interrelationship mining to real-life data.

A Proofs of Theoretical Results

Lemma 5.1 *The collection S of sets defined as*

$$S = \left\{ \begin{array}{l} R(\mathsf{a},\mathsf{b}),\ R^c(\mathsf{a},\mathsf{b}),\ LN(\mathsf{a},\mathsf{b}), \\ RN(\mathsf{a},\mathsf{b}),\ BN(\mathsf{a},\mathsf{b}) \end{array} \right\} - \{\emptyset\}$$

comprises a partition of U.

Proof From the definitions of support set $R(\mathsf{a},\mathsf{b})$, nonsupport set $R^c(\mathsf{a},\mathsf{b})$, and $LN(\mathsf{a},\mathsf{b})$, $RN(\mathsf{a},\mathsf{b})$, and $BN(\mathsf{a},\mathsf{b})$, it is obvious that, for any $S_i, S_j \in S$, if $S_i \neq \emptyset$, $S_j \neq \emptyset$, and $i \neq j$ hold, then $S_i \cap S_j = \emptyset$ holds. Thus, it is sufficient to show $\cup_{S \in S} S = U$. Moreover, every set $S \in S$ is obviously a subset of U, which implies $\cup_{S \in S} S \subseteq U$.

Let $x \in U$ be any object. If both $v = \rho(x,\mathsf{a}) \neq *$ and $w = \rho(x,\mathsf{b}) \neq *$ hold, either $(v,w) \in R$ or $(v,w) \notin R$ holds, which implies that either $x \in R(\mathsf{a},\mathsf{b})$ or $x \in R^c(\mathsf{a},\mathsf{b})$ holds. Otherwise, if $v = *$ or $w = *$ or both hold, either $x \in LN(\mathsf{a},\mathsf{b})$ or $x \in RN(\mathsf{a},\mathsf{b})$ or $x \in BN(\mathsf{a},\mathsf{b})$ hold. These facts imply that $\cup_{S \in S} S \supseteq U$; thus, $\cup_{S \in S} S = U$.

Theorem 5.1 *Let $\mathsf{a}, \mathsf{b} \in C$ be two condition attributes in a given decision table DT and $R \subseteq V_\mathsf{a} \times V_\mathsf{b}$ be a binary relation. The binary relation $SIM(\mathsf{a}R\mathsf{b})$ defined by (5.24) satisfies reflexivity and symmetry.*

Proof First, we show that $SIM(\mathsf{a}R\mathsf{b})$ satisfies reflexivity, i.e., $(x,x) \in SIM(\mathsf{a}R\mathsf{b})$ for all $x \in U$. Because S is a partition by Lemma 5.1, for every $x \in U$, there exists a set $S \in S$ such that $x \in S$; thus $(x,x) \in SIM(\mathsf{a}R\mathsf{b})$ for all $x \in U$ by (5.24).

Next, we show symmetry, i.e., if $(x, y) \in SIM(\mathsf{a}R\mathsf{b})$ holds, then $(y, x) \in SIM(\mathsf{a}R\mathsf{b})$ holds. Suppose $(x, y) \in SIM(\mathsf{a}R\mathsf{b})$ holds. If there exists a set $S \in \mathcal{S}$ such that $x, y \in S$ holds, it is obvious that $(y, x) \in SIM(\mathsf{a}R\mathsf{b})$ also holds. We then assume that $x \in NULL(\mathsf{a}, \mathsf{b})$, which implies that $y \notin NULL(\mathsf{a}, \mathsf{b})$ by (5.24). Therefore, $(y, x) \in SIM(\mathsf{a}R\mathsf{b})$ also holds by (5.24). The case of $x \notin NULL(\mathsf{a}, \mathsf{b})$ is shown similarly, which indicates that $SIM(\mathsf{a}R\mathsf{b})$ satisfies symmetry.

Theorem 5.2 *Let DT_{int} be a decision table for interrelationship mining defined by (5.12) and assume that DT_{int} has a revised information function ρ_{int} defined by (5.26). For any interrelated condition attribute $\mathsf{a}R\mathsf{b} \in AT_{int}$, the binary relation $SIM(\{\mathsf{a}R\mathsf{b}\})$ defined by (5.27) satisfies reflexivity and symmetry.*

Proof It is obvious that $\rho(x, \mathsf{a}R\mathsf{b}) = \rho(x, \mathsf{a}R\mathsf{b})$ for every $x \in U$; therefore, $(x, x) \in SIM(\{\mathsf{a}R\mathsf{b}\})$ holds for every $x \in U$, i.e., the binary relation $SIM(\{\mathsf{a}R\mathsf{b}\})$ satisfies reflexivity.

We show that $SIM(\{\mathsf{a}R\mathsf{b}\})$ satisfies symmetry. Suppose $(x, y) \in SIM(\{\mathsf{a}R\mathsf{b}\})$ holds. If $\rho(x, \mathsf{a}R\mathsf{b}) = \rho(y, \mathsf{a}R\mathsf{b})$ holds, it is obvious that $\rho(y, \mathsf{a}R\mathsf{b}) = \rho(x, \mathsf{a}R\mathsf{b})$ also holds and $(y, x) \in SIM(\{\mathsf{a}R\mathsf{b}\})$. If $\rho(x, \mathsf{a}R\mathsf{b}) \neq \rho(y, \mathsf{a}R\mathsf{b})$ and $\rho(x, \mathsf{a}R\mathsf{b}) \in \{1, 0\}$ hold, according to the definition of $SIM(\{\mathsf{a}R\mathsf{b}\})$ by (5.27), it is implied that $\rho(y, \mathsf{a}R\mathsf{b}) \in \{*_l, *_r, *_b\}$ holds. Thus, $(y, x) \in SIM(\{\mathsf{a}R\mathsf{b}\})$. We can also prove the case of $\rho(x, \mathsf{a}R\mathsf{b}) \in \{*_l, *_r, *_b\}$ similarly. Thus, $SIM(\{\mathsf{a}R\mathsf{b}\})$ satisfies symmetry.

Theorem 5.3 *Let DT be a decision table by (5.1) and DT_{int} be a decision table for interrelationship mining by (5.12) with the redefined information function by (5.26). The following equality holds.*

$$SIM_{DT}(\mathsf{a}R\mathsf{b}) = SIM_{DT_{int}}(\{\mathsf{a}R\mathsf{b}\}),$$

where $SIM_{DT}(\mathsf{a}R\mathsf{b})$ is the similarity relation in DT with respect to the interrelationship between a and b by R defined by (5.24), and $SIM_{DT_{int}}(\{\mathsf{a}R\mathsf{b}\})$ is the similarity relation in DT_{int} constructed from the singleton $\{\mathsf{a}R\mathsf{b}\}$ of the interrelated condition attribute $\mathsf{a}R\mathsf{b}$.

Proof Suppose that $(x, y) \in SIM_{DT}(\mathsf{a}R\mathsf{b})$ holds. If there exists a set $S \in \mathcal{S}$ such that $x, y \in S$ holds, it is clear that $\rho_{int}(x, \mathsf{a}R\mathsf{b}) = \rho_{int}(y, \mathsf{a}R\mathsf{b})$ holds by the definition of the redefined information function ρ_{int} by (5.26), which indicates that $(x, y) \in SIM_{DT_{int}}(\{\mathsf{a}R\mathsf{b}\})$ by (5.27). Otherwise, if $x \in NULL(\mathsf{a}, \mathsf{b})$ and $y \notin NULL(\mathsf{a}, \mathsf{b})$, it is implied that $\rho_{int}(x, \mathsf{a}R\mathsf{b}) \in \{*_l, *_r, *_b\}$ and $\rho_{int}(y, \mathsf{a}R\mathsf{b}) \in \{0, 1\}$ hold by (5.26); thus, $(x, y) \in SIM_{DT_{int}}(\{\mathsf{a}R\mathsf{b}\})$ holds by (5.27). We can also show the case of $x \in NULL(\mathsf{a}, \mathsf{b})$ and $y \notin NULL(\mathsf{a}, \mathsf{b})$ similarly, thus, $SIM_{DT}(\mathsf{a}R\mathsf{b}) \subseteq SIM_{DT_{int}}(\{\mathsf{a}R\mathsf{b}\})$ holds. The opposite inclusion $SIM_{DT}(\mathsf{a}R\mathsf{b}) \supseteq SIM_{DT_{int}}(\{\mathsf{a}R\mathsf{b}\})$ is also provable in the same manner as the above proof, which completes the proof.

References

1. Greco, S., Matarazzo, B., Słowiński, R.: Rough set theory for multicriteria decision analysis. Eur. J. Oper. Res. **129**, 1–47 (2002)
2. Kryszkiewicz, M.: Rough set approach to incomplete information systems. Inf. Sci. **112**, 39–49 (1998)
3. Kudo, Y., Murai, T.: A note on application of interrelationship mining to incomplete information systems. In: Proceedings of the 31st Fuzzy System Symposium, pp. 777–778 (2015) (in Japanese)
4. Kudo, Y., Murai, T.: A plan of interrelationship mining using rough sets. In: Proceedings of the 29th Fuzzy System Symposium, pp. 33–36 (2013) (in Japanese)
5. Kudo, Y., Murai, T.: Decision logic for rough set-based interrelationship mining. In: Proceedings of IEEE GrC 2013, pp. 172–177 (2013)
6. Kudo, Y., Murai, T.: Indiscernibility relations by interrelationships between attributes in rough set data analysis. In: Proceedings of IEEE GrC 2012, pp. 264–269 (2012)
7. Kudo, Y., Murai, T.: Interrelationship mining from a viewpoint of rough sets on two universes. In: Proceedings of IEEE GrC 2014, pp. 137–140 (2014)
8. Kudo, Y., Murai, T.: On representation ability of interrelated attributes in rough set-based Interrelationship mining. In: Proceedings of ISIS 2015, pp. 1229–1237 (2015)
9. Kudo, Y., Murai, T.: Some properties of interrelated attributes in relative reducts for interrelationship mining, In: Proceedings of SCIS&ISIS 2014, SOFT, pp. 998–1001 (2014)
10. Kudo, Y., Okada, Y., Murai, T.: On a possibility of applying interrelationship mining to gene expression data analysis. Brain and Health Informatics. LNAI, vol. 8211, 379–388 (2013)
11. Kudo, Y., Murai, T.: An evaluation method of relative reducts based on roughness of partitions. Int. J. Cogn. Inform. Nat. Intell. **4**, 50–62 (2010)
12. Pawlak, Z.: Rough sets. Int. J. Comput. Inf. Sci. **11**, 341–356 (1982)
13. Pawlak, Z.: Rough Sets: Theoretical Aspects of Reasoning about Data. Kluwer, Dordrecht (1991)
14. Pawlak, Z., Skowron, A.: Rough sets: some extensions. Inf. Sci. **177**, 28–40 (2007)
15. Yao, Y.Y., Zhou, B., Chen, Y.: Interpreting low and high order rules: a granular computing approach. In: Proceedings of RSEISP 2007. LNCS, vol. 4585, pp. 371–380 (2007)

Chapter 6
A Parallel Computation Method for Heuristic Attribute Reduction Using Reduced Decision Tables

Abstract This chapter proposes a parallel computation framework for a heuristic attribute reduction method. Attribute reduction is a key technique to use rough set theory as a tool in data mining. The authors have previously proposed a heuristic attribute reduction method to compute as many relative reducts as possible from a given dataset with numerous attributes. We parallelize our method by using open multiprocessing. We also evaluate the performance of a parallelized attribute reduction method by experiments.

Keywords Rough set · Attribute reduction · Parallel computation · Open multiprocessing

6.1 Introduction

Rough set theory, proposed by Pawlak [8, 9], presents mathematical foundations for approximation of concepts and reasoning about data. *Attribute reduction* to compute relative reducts from a given dataset is a key technique to use rough set theory as a tool in data mining. However, the exhaustive attribute reduction problem, i.e., extraction of all relative reducts from a given dataset, is NP-hard, and therefore, various heuristic algorithms have been proposed to compute a small number of relative reducts [2–4, 6, 16]. The authors have previously proposed a heuristic attribute reduction method to compute as many relative reducts as possible from a decision table with numerous condition attributes [5].

In this chapter, we introduce a parallel computation framework for the authors' method to improve the computation time of attribute reduction. We also evaluate the performance of a parallelized heuristic attribute reduction method by experiments.

This chapter is organized as follows. In Sect. 6.2, we review Pawlak's rough set theory and the authors' heuristic attribute reduction method to compute as many relative reducts as possible from a decision table with numerous condition attributes [5]. In Sect. 6.3, we introduce a parallel computation framework for our method. We apply the parallel computation framework to five datasets in Sect. 6.4 and discuss the effects of parallel computation in Sect. 6.5. Section 6.6 presents conclusions. Note that this paper is a revised and expanded version of the original manuscript [7].

© Springer Nature Switzerland AG 2020
S. Akama et al., *Topics in Rough Set Theory*, Intelligent Systems
Reference Library 168, https://doi.org/10.1007/978-3-030-29566-0_6

6.2   Rough Sets

In this section, we review rough set theory as backgrounds for this paper. The contents of this section are based on [9, 10].

6.2.1   Decision Tables and Indiscernibility Relations

Generally, data analysis subjects by rough sets are described by decision tables. Formally, a decision table is characterized by the following triple:

$$DT = (U, C, \mathsf{d}), \tag{6.1}$$

where U is a finite and nonempty set of objects, C is a finite and nonempty set of condition attributes, and d is a decision attribute such that $\mathsf{d} \notin C$. Each attribute $\mathsf{a} \in C \cup \{\mathsf{d}\}$ is a function $\mathsf{a} : U \to V_\mathsf{a}$, where V_a is a set of values of the attribute a.

On the basis of subsets of attributes, indiscernibility relations provide classification of objects in decision tables. For any set of attributes $A \subseteq C \cup \{\mathsf{d}\}$, the indiscernibility relation $IND(A)$ is the following binary relation on U:

$$IND(A) = \{(x, y) \mid \mathsf{a}(x) = \mathsf{a}(y), \forall \mathsf{a} \in A\}. \tag{6.2}$$

If a pair (x, y) is in $IND(A)$, then two objects x and y are indiscernible with respect to all attributes in A. The equivalence class $[x]_B$ of $x \in U$ under $IND(B)$ is the set of objects that is not discernible from x even though all attributes in B are used. Any indiscernibility relation $IND(B)$ provides a partition of U, i.e., the quotient set $U/IND(B)$. In particular, the partition $\mathcal{D} = \{D_1, \ldots, D_m\}$ provided by the indiscernibility relation $IND(\{\mathsf{d}\})$, on the basis of decision attribute d, is called the set of decision classes.

Classifying objects with respect to condition attributes provides an approximation of the decision classes. Formally, for any set $B \subseteq C$ of condition attributes and any decision class $D_i \in \mathcal{D}$, we let:

$$\underline{B}(D_i) = \{x \in U \mid [x]_B \subseteq D_i\}. \tag{6.3}$$

The set $\underline{B}(D_i)$ is called a lower approximation of the decision class D_i with respect to B. The lower approximation $\underline{B}(D_i)$ is the set of objects that is correctly classified with respect to the decision class D_i by checking all attributes in B.

Table 6.1 is an example of a decision table that consists of the following objects: $U = \{x_1, \ldots, x_7\}, C = \{\mathsf{c}_1, \ldots, \mathsf{c}_8\}$, and d. The decision attribute d provides the following three decision classes: $D_1 = \{x_1, x_5\}, D_2 = \{x_3, x_4, x_7\}.$ and $D_3 = \{x_2, x_6\}$.

Table 6.1 Decision table

U	c_1	c_2	c_3	c_4	c_5	c_6	c_7	c_8	d
x_1	1	1	1	1	1	1	1	2	1
x_2	3	1	3	2	2	1	1	2	3
x_3	2	3	2	1	2	2	1	1	2
x_4	2	2	2	1	2	2	3	1	2
x_5	2	2	3	1	1	1	1	2	1
x_6	3	2	1	2	2	1	2	2	3
x_7	1	1	1	1	1	1	1	1	2

6.2.2 Relative Reducts

By checking values of all condition attributes, we can classify all discernible objects of the given decision table with respect to the corresponding decision classes. However, not all condition attributes may need to be checked in the sense that some condition attributes are essential to classify and other attributes are redundant. A minimal set of condition attributes to classify all discernible objects to correct decision classes is called a relative reduct of the decision table.

For any subset $X \subseteq C$ of condition attributes in a decision table DT, we let:

$$POS_X(\mathcal{D}) = \bigcup_{D_i \in \mathcal{D}} \underline{X}(D_i). \tag{6.4}$$

The set $POS_X(\mathcal{D})$ is called the positive region of $\mathcal{D}$ by X. All objects $x \in POS_X(\mathcal{D})$ are classified into correct decision classes by checking all attributes in X. In particular, the set $POS_C(\mathcal{D})$ is the set of all discernible objects in the given decision table.

Here, we formally define relative reducts. A set $A \subseteq C$ is called a relative reduct of the decision table DT if the set A satisfies the following conditions:

1. $POS_A(\mathcal{D}) = POS_C(\mathcal{D})$.
2. $POS_B(\mathcal{D}) \neq POS_C(\mathcal{D})$ for any proper subset $B \subset A$.

In general, there are plural relative reducts in a decision table. The common part of all relative reducts is called the core of the decision table. For example, in Table 6.1, there are four relative reducts: $\{c_1, c_8\}$, $\{c_4, c_8\}$, $\{c_5, c_8\}$, and $\{c_2, c_3, c_8\}$. The condition attribute c_8 appears in all relative reducts in Table 6.1 and therefore the core of Table 6.1 is $\{c_8\}$.

6.2.3 Discernibility Matrix

The *discernibility matrix* [11] is one of the most popular methods to compute all relative reducts in the decision table. Let DT be a decision table with $|U|$ objects, where $|U|$ is the cardinality of U. The discernibility matrix DM of DT is a symmetric $|U| \times |U|$ matrix whose element at the i-th row and j-th column is the following set of condition attributes to discern between two objects x_i and x_j:

$$
\delta_{ij} =
\begin{cases}
\{\mathsf{a} \in C \mid \mathsf{a}(x_i) \neq \mathsf{a}(x_j)\}, \\
\quad \text{if } \mathsf{d}(x_i) \neq \mathsf{d}(x_j) \text{ and } \{x_i, x_j\} \cap POS_C(\mathcal{D}) \neq \emptyset, \\
\emptyset, \qquad\qquad\qquad \text{otherwise.}
\end{cases}
\tag{6.5}
$$

Each element $\mathsf{a} \in \delta_{ij}$ represents that the objects x_i and x_j are discernible by checking the value of the attribute a.

However, the problem of computing all relative reducts is an NP-hard problem [11].

6.2.4 Heuristic Attribute Reduction Using Reduced Decision Tables

In this subsection, we review a heuristic method to generate as many relative reducts as possible from decision tables with numerous condition attributes proposed by Kudo and Murai [5].

This heuristic method is based on the concept of reduced decision tables that preserve the discernibility of objects that belong to different decision classes in the given decision table. Formally, a reduced decision table of a given decision table is defined as follows.

Definition 6.1 Let $DT = (U, C, \mathsf{d})$ be a decision table. A reduced decision table of DT is the following triple:

$$
RDT = (U, C', \mathsf{d}),
\tag{6.6}
$$

where U and d are identical to DT. The set of condition attributes C' satisfies the following conditions:

1. $C' \subseteq C$.
2. For any objects x_i and x_j that belong to different decision classes, if x_i and x_j are discernible by $IND(C)$, x_i and x_j are also discernible by $IND(C')$.

Algorithm 2 generates a reduced decision table of the given decision table. In Algorithm 2, condition attributes are randomly selected from C on the basis of the parameter of base size, b, that decides the minimum number of condition attributes of the reduced decision table and supply some attributes in elements of the discernibility matrix to preserve discernibility of objects in the given decision table.

Algorithm 2 dtr: decision table reduction algorithm

Input: decision table $DT = (U, C, \mathsf{d})$, discernibility matrix DM of DT, base size b
Output: reduced decision table (U, C', d)
1: Select b attributes $\mathsf{a}_1, \cdots, \mathsf{a}_b$ from C at random by sampling without replacement
2: $C' = \{\mathsf{a}_1, \cdots, \mathsf{a}_b\}$
3: **for all** $\delta_{ij} \in DM$ such that $i > j$ **do**
4: **if** $\delta_{ij} \neq \emptyset$ **and** $\delta_{ij} \cap C' = \emptyset$ **then**
5:    Select $\mathsf{c} \in \delta_{ij}$ at random
6:    $C' = C' \cup \{\mathsf{c}\}$
7: **end if**
8: **end for**
9: **return** (U, C', d)

Note that for any decision table $DT = (U, C, \mathsf{d})$ and any reduced decision table $RDT = (U, C', \mathsf{d})$ of the decision table DT, a set of condition attributes $A \subseteq C'$ is a relative reduct of the reduced decision table RDT if and only if A is also a relative reduct of the given decision table DT [5]. Thus generating as many reduced decision tables and their relative reducts as possible from the given decision table, we can extract many relative reducts from the given decision table.

Algorithm 3 generates relative reducts of the given decision table on the basis of generating reduced decision tables using Algorithm 2 and applying exhaustive attribute reduction and heuristic attribute reduction according to the number of condition attributes of each reduced decision table.

Algorithm 3 Exhaustive / heuristic attribute reduction

Input: decision table $DT = (U, C, d)$,
    base size b, size limit L, number of iteration I
Output: set of candidates of relative reduct RED
1: $RED = \emptyset$
2: $DM \leftarrow$ the discernibility matrix of DT
3: **if** $|C| \leq L$ **then**
4: $RED \leftarrow$ result of exhaustive attribute reduction from DT
5: **else**
6: **for** $i = 1$ to I **do**
7:    $RDT = dtr(DT, DM, b)$
8:    **if** $|C'| \leq L$ **then**
9:      $S \leftarrow$ result of exhaustive attribute reduction from RDT
10:   **else**
11:     $S \leftarrow$ result of heuristic attribute reduction from RDT
12:   **end if**
13:   $RED = RED \cup S$
14: **end for**
15: **end if**
16: **return** RED

In Algorithm 3, the size limit, L, is the threshold for switching attribute reduction methods and if the number of condition attributes of a decision table is smaller than L, Algorithm 3 tries to generate the set of all relative reducts of the decision table. Thus, we need to appropriately set the threshold L. If the number of condition attributes of the given decision table, DT, is greater than the threshold L, Algorithm 3 iterates I times the generation of a reduced decision table, RDT, and attribute reduction from RDT by selecting the exhaustive method or the heuristic method and generates the set RED of relative reducts. Note that RED may contain some candidate with redundant condition attributes if the result of the heuristic attribute reduction is not guaranteed to generate relative reducts.

6.3   Parallel Computation of Heuristic Attribute Reduction

In this section, we introduce a parallel computation framework for the authors' heuristic attribute reduction method using reduced decision tables.

6.3.1   OpenMP

We used *open multiprocessing* (OpenMP) [12] to parallelize the authors' heuristic attribute reduction method in Sect. 6.2.4. In this subsection, we review OpenMP on the basis of the contents in [12, 13].

OpenMP is an application program interface (API) that supports multi platform shared memory multi processing programing in C/C++ and Fortran on many architectures. Parallelization mechanism implemented in OpenMP is based on fork-join model of multi-threading; a master thread calls a specified number of slave threads, and a task is divided among them. Each thread independently executes the parallelized section of code and joins back to the master thread.

6.3.2   Parallelization of Heuristic Attribute Reduction

To parallelize heuristic attribute reduction using OpenMP, we divided for-loops between steps 6 and 14 in Algorithm 3 into a specified number of slave threads. Each pass through the loop consists of construction of a reduced decision table and extraction of relative reducts by exhaustive/heuristic attribute reduction from the reduced decision table. Computation of each pass does not use results of other passes, and therefore each pass can be independently processed.

We implemented the heuristic attribute reduction method in Sect. 6.2.4 that was parallelized by OpenMP in C++ on a Linux server with CPU: Intel Xeon X5690 3.47GHz (6cores) × 2, Memory: 96GB, OS: CentOS release 5.7, and HDD: 1TB, and compiled by gcc 4.4.4. In the implementation of Algorithm 3, we used the discernibility matrix for the exhaustive attribute reduction and a heuristic algorithm to generate one relative reduct candidate [16] as the heuristic attribute reduction.

6.4 Experiments

In this section, we evaluate the performance of the parallelized heuristic attribute reduction method.

6.4.1 Methods

The following five datasets were used for experiments; audiology, lung cancer, internet advertisements [14], breast cancer [15], and leukemia datasets [1]. Table 6.2 describes the number of samples, condition attributes, and decision classes of these five datasets.

To evaluate the effect on computation time reduction by increasing slave threads to independently compute for-loops in Algorithm 3, we applied our method 50 times to each dataset with twelve cases; Only one (serial computation) to 12 threads were used (fully parallelized computation that was enabled in the used Linux server, i.e., each core in the CPUs was used for one thread), and the average computation time of each case was measured. Table 6.3 describes the parameters used in Algorithms 2 and 3. We determined these parameters on the basis of results of pre-experiments.

6.4.2 Experiment Results

Table 6.5 shows the average computation times of the five datasets using 1~12 threads. Each row describes the average computation time (s) and the standard deviation for each dataset, and each column shows the number of threads used for com-

Table 6.2 Datasets used in experiments

Name	Samples	Cond. Att.	Dec. Cls.
Audiology (Standardized)	226	69	24
Breast cancer	49	7129	2
Internet advertisements	3279	1558	2
Leukemia	52	12582	2
Lung cancer	32	56	3

Table 6.3 Parameters used in experiments

Parameter	Value
Base size b	15
Size limitation L	30
Number of iteration I	500

Fig. 6.1 Time reduction ratios of each dataset with each number of threads

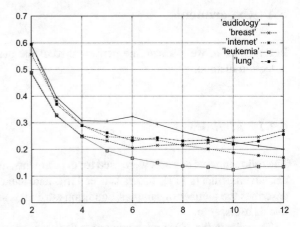

Table 6.4 Best TRR of each dataset

Name	Best TRR	# of threads
Audiology	0.201	12
Breast	0.204	6
Internet	0.170	12
Leukemia	0.124	10
Lung	0.219	10

putation. The result in bold letters is the fastest average computation time of each dataset. For example, the value **13.13***±0.48 in breast cancer dataset in column 6 denotes the average computation time. The average computation time for over 50 applications of our method to breast cancer dataset using six threads was 13.13 s and the standard deviation was 0.48 s. There were statistically significant differences between the average computation time with six threads and all other average computation times, and therefore, the average computation time with six threads was the fastest average computation time on breast cancer dataset.

Table 6.6 shows the average number of times that exhaustive attribute reduction was used in $I = 500$ iterations with each number of threads. In the breast cancer, leukemia, and lung cancer datasets, every attribute reduction in all numbers of threads was performed by exhaustive attribute reduction.

On the other hand, in the internet advertisement dataset, every attribute reduction was performed by heuristic attribute reduction. In the audiology dataset, both exhaustive and heuristic attribute reduction were used. Because Algorithm 3 switches between exhaustive attribute reduction and heuristic attribute reduction depending on the number of condition attributes in the given reduced decision table, simple comparison of computation time between datasets is difficult.

We then compared time reduction ratios (TRRs) with respect to the number of threads. Figure 6.1 shows TRRs of five datasets. The horizontal axis describes the number of threads, and the vertical axis describes the TRR by using n threads that is defined by

Table 6.5 Average computation time(s) of each dataset using 1~12 threads

| Name | Number of threads | | | | | |
	1	2	3	4	5	6
Audiology	3.89±0.01	2.30±0.02	1.54±0.02	1.20±0.07	1.19± 0.24	1.26±0.14
Breast	64.19±1.09	31.17±0.42	20.86±0.31	16.09±0.24	14.87±0.43	**13.13***±0.48
Internet	371.25±2.39	206.70±5.22	141.24±4.59	107.41±3.90	91.69±7.57	90.36±11.41
Leukemia	40.02±0.68	19.54±0.34	13.14±0.23	9.91±0.16	7.77±0.22	6.67±0.16
Lung	8.15±0.58	4.85±0.25	3.01±0.12	2.35±0.15	2.13±0.17	1.90±0.12

| Name | Number of threads | | | | | |
	7	8	9	10	11	12
Audiology	1.14±0.06	1.03±0.04	0.95±0.02	0.88±0.03	0.83±0.02	**0.78***±0.02
Breast	13.79±0.61	13.95±0.66	14.37±0.67	15.68±0.70	15.81±0.77	17.31±0.97
Internet	87.32±5.42	79.57±4.45	74.80±1.35	69.50±2.36	65.71±1.26	**62.93***±1.05
Leukemia	5.99±0.18	5.49±0.16	5.30±0.17	**4.95***±0.20	5.42±0.15	5.40±0.15
Lung	1.98±0.13	1.89±0.09	1.91±0.10	**1.79***±0.12	1.88±0.11	2.09±0.20

$^*p < 0.01$

Table 6.6 Average number of application of exhaustive attribute reduction in 500 iterations

| Name | Number of threads | | | | | | | | | | | |
	1	2	3	4	5	6	7	8	9	10	11	12
Audiology	43.0	43.6	41.2	41.6	44.6	39.7	41.5	42.4	38.4	43.3	46.5	45.5
Breast	500	500	500	500	500	500	500	500	500	500	500	500
Internet	0	0	0	0	0	0	0	0	0	0	0	0
Leukemia	500	500	500	500	500	500	500	500	500	500	500	500
Lung	500	500	500	500	500	500	500	500	500	500	500	500

$$TRR(n) = \frac{\text{Computation time with } n \text{ threads}}{\text{Computation time with one thread}}. \tag{6.7}$$

By this definition, $TRR(1) = 1$ holds for all datasets. Table 6.4 describes the best TRR and the number of threads that performed the best for each dataset.

6.5 Discussion

For any fixed number of threads, the order of the length of computation time of the datasets was fixed; audiology < lung < leukemia < breast < internet. In particular, the computation time of the internet advertisement dataset was very long than that of the other datasets. We consider that the number of samples in the internet advertisement dataset caused this long computation time. In this paper, because both the exhaustive/heuristic attribute reduction used in Algorithm 3 require a discernibility matrix of a decision table, in each iteration in Algorithm 3, we need to construct a discernibility matrix of a reduced decision table. The computational complexity of constructing a discernibility matrix from a decision table $DT = (U, C, d)$ is $O(|U|^2|C|)$ [11], and therefore the construction of a discernibility matrix with $|U| = 3279$ from a reduced decision table $RDT = (U, C', d)$ in each iteration severely affected the computation time.

In general, for each dataset, increasing the number of threads tends to decrease the average computation time. For example, by using two threads, the average computation times decreased to around 49–59% of the case of serial computation. However, the decrease of TRRs becomes slow as the number of threads increases. In the breast cancer, lung cancer, and leukemia datasets, the TRR began to increase at some number of threads. We think that this is because the overhead of parallelization exceeded the effect of time reduction by parallelization. Each thread requires new memory space in the shared memory space to store the result of attribute reduction. As shown in Table 6.6, exhaustive attribute reduction was used for the breast cancer, leukemia, and lung cancer datasets. For these three datasets, computation of all relative reducts by exhaustive attribute reduction in each thread was easy and very fast because the reduced decision table was sufficiently small. Consequently, if the number of threads is excessive to compute attribute reduction of the given data set, memory allocation requirements by threads occur too frequently and this causes the bottleneck of computation time reduction.

6.6   Conclusion and Future Issues

In this chapterr, we introduced OpenMP to a heuristic attribute reduction method that was originally proposed by the authors [5] to parallelize the computation of attribute reduction. The experiment results indicate that using an appropriate number of threads is effective to parallelize the heuristic attribute reduction method.

To improve the computation time of the authors' heuristic attribute reduction method, many issues are unsolved. One of the most important issues is to improve the computation time of each loop in Algorithm 3, i.e., construction of a reduced decision table and exhaustive/heuristic attribute reduction of the reduced decision table.

In this chapter, we restricted ourselves to the one parameters setting shown in Table 6.3. We plan to investigate the effects of parameters setting on computation times. We also need to consider how to set appropriate parameters for each dataset. Adaptive control of the number of threads to determine an appropriate number of threads for each dataset is also an interesting future issue.

References

1. Armstrong, S.A., et al.: MLL translocations specify a distinct gene expression profile that distinguishes a unique leukemia. Nat. Genet. **30**, 41–47 (2002)
2. Chouchoulas, A., Shen, A.: Rough set-aided keyword reduction for text categorization. Appl. Artif. Intell. **15**, 843–873 (2001)

3. Hu, K. Diao, L., Lu, Y., Shi, C.: A heuristic optimal reduct algorithm. In: Proceedings of IDEAL 2000. LNSC, vol. 1983, pp. 139–144. Springer (2000)

4. Kryszkiewicz, M. and Lasek, P.: FUN: Fast discovery of minimal sets of attributes functionally determining a decision attribute. Transactions on Rough Sets IX. LNCS, vol. 5390, pp. 76–95. Springer, Berlin (2008)

5. Kudo, Y., Murai, T.: An attribute reduction algorithm by switching exhaustive and heuristic computation of relative reducts. In: Proceedings of IEEE GrC2010, pp. 265–270. IEEE (2010)

6. Kudo, Y., Murai, T.: Heuristic algorithm for attribute reduction based on classification ability by condition attributes. J. Adv. Comput. Intell. Intell. Inform. **15**, 102–109 (2011)

7. Kudo, Y., Murai, T.: A parallel computation method of attribute reduction. In: Proceedings of ISCIIA 2012 (2012)

8. Pawlak, Z.: Rough sets. Int. J. Comput. Inf. Sci. **11**, 341–356 (1982)

9. Pawlak, Z.: Rough Sets: Theoretical Aspects of Reasoning about Data. Kluwer, Dordrechtr (1991)

10. Polkowski, L.: Rough Sets: Mathematical Foundations. Advances in Soft Computing. Physica-Verlag, Heidelberg (2002)

11. Skowron, A., Rauszer, C.M.: The discernibility matrix and functions in information systems. In: Słowiński, R. (ed.) Intelligent Decision Support: Handbook of Application and Advance of the Rough Set Theory, pp. 331–362. Kluwer, Dordrecht (1992)

12. OpenMP; http://openmp.org/wp/

13. OpenMP from Wikipedia; http://en.wikipedia.org/wiki/OpenMP/

14. UCI Machine Learning Repository; http://archive.ics.uci.edu/ml/

15. West, M., et al.: Predicting the clinical status of human breast cancer by using gene expression profiles. Proc. Natl. Acad. Sci. **98**, 11462–11467 (2001)

16. Zhang, J., Wang, J., Li, D., He, H., Sun, J.: A new heuristic reduct algorithm based on rough sets theory. In: Proceedings of WAIM 2003. LNCS, vol. 2762, pp. 247–253. Springer (2003)

Chapter 7
Heuristic Algorithm for Attribute Reduction Based on Classification Ability by Condition Attributes

Abstract This chapter discusses the heuristic algorithm to computes a relative reduct candidate based on evaluating classification ability of condition attributes. Considering the discernibility and equivalence of objects for condition attributes in relative reducts, we introduce evaluation criteria for condition attributes and relative reducts. The computational complexity of the proposed algorithm is $O(|U|^2|C|^2)$. Experimental results indicate that our algorithm often generates a relative reduct producing a good evaluation result.

Keywords Rough set · Attribute reduction · Heuristic algorithm · Classification ability

7.1 Introduction

Focusing on generating relative reducts from a given decision table, Pawlak's *rough set theory* [11, 13] provides a theoretical data-mining framework for categorical data. Among related studies, Skowron and Rauszer [16] proposed an algorithm to compute all relative reducts using the discernibility matrix (DM) and proved that the computational complexity of computing all relative reducts in a given decision table is NP-hard.

Most of heuristic algorithms proposed to compute relative reducts instead of computing all relative reducts [1–8, 14, 17–19, 21–23] use certain criteria for selecting condition attributes. However, quality of relative reducts as outputs of these algorithms are generally not evaluated except for the length, i.e., the number of condition attributes in relative reduct. Introducing a relative reduct evaluation criterion based on discernibility and equivalence by condition attributes, we propose a heuristic algorithm using our criterion to compute a relative reduct candidate giving a good evaluation result.

The rest of this chapter is organized as follows. In Sect. 7.2, we review Pawlak's rough set theory. In Sect. 7.3, we introduce evaluation criteria for classification ability of condition attributes and relative reducts and consider theoretical aspects of these criteria. In Sect. 7.4, we propose a heuristic algorithm using our criteria to compute

S. Akama et al., *Topics in Rough Set Theory*, Intelligent Systems
Reference Library 168, https://doi.org/10.1007/978-3-030-29566-0_7

a relative reduct candidate giving good evaluation result. We apply our algorithm to UCI repository datasets [24] in Sect. 7.5 and discuss the algorithm's computational complexity in Sect. 7.6. Section 7.7 presents conclusions. Note that this chapter is a revised, expanded version of the original manuscript [9].

7.2   Rough Set Theory

We begin by reviewing rough set theory, focusing on decision tables, relative reducts, and DMs. Note that this review is based on [10, 15].

7.2.1   Decision Tables and Lower and Upper Approximations

Decision tables are generally used to describe data analysis subjects in rough sets. Formally, a decision table is the following quadruple:

$$DT = (U, C \cup D, V, \rho), \tag{7.1}$$

where U is a finite, nonempty set of objects, C and D are finite and nonempty sets of condition attributes and decision attributes such that $C \cap D = \emptyset$, V is a set of all values of attributes $a \in C \cup D$, and $\rho : U \times (C \cup D) \to V$ is a function that assigns value $\rho(x, a) \in V$ at attribute $a \in C \cup D$ to object $x \in U$.

Indiscernibility relations based on attribute subsets classify objects in decision tables. For any set of attributes $A \subseteq C \cup D$, indiscernibility relation R_A is the following binary relation on U:

$$R_A = \{(x, y) \mid \rho(x, a) = \rho(y, a), \forall a \in A\}. \tag{7.2}$$

If pair (x, y) is in R_A, then two objects x and y are indiscernible for all attributes in A. Any indiscernibility relation is an equivalence relation and equivalence classes by an equivalence relation consist of a partition on the domain of the equivalence relation. Specifically, indiscernibility relation R_D based on set of decision attributes D provides partition $\mathcal{D} = \{D_1, \ldots, D_k\}$, and each element $D_i \in \mathcal{D}$ is called a decision class.

Classifying objects for condition attributes provides an approximation of decision classes. Formally, for any set $B \subseteq C$ of condition attributes and any decision class $D_i \in \mathcal{D}$, we let:

$$\underline{B}(D_i) = \{x \in U \mid [x]_B \subseteq D_i\}, \tag{7.3}$$
$$\overline{B}(D_i) = \{x \in U \mid [x]_B \cap D_i \neq \emptyset\}, \tag{7.4}$$

Table 7.1 Decision table

U	c_1	c_2	c_3	c_4	c_5	c_6	d
x_1	1	0	0	0	0	1	1
x_2	0	1	0	0	0	1	1
x_3	0	2	1	0	0	0	2
x_4	0	1	1	1	1	0	2
x_5	0	1	2	0	0	1	1
x_6	0	1	0	0	1	1	3

where set $[x]_B$ is the equivalence class of x by indiscernibility relation R_B. Set $\underline{B}(D_i)$ is called a lower approximation and set $\overline{B}(D_i)$ is called an upper approximation of decision class D_i with respect to B. Note that lower approximation $\underline{B}(D_i)$ is the set of objects correctly classified into the decision class D_i by checking all attributes in B. A decision table is called consistent if and only if $\underline{C}(D_i) = D_i = \overline{C}(D_i)$ holds for all decision classes $D_i \in \mathcal{D}$.

Example 7.1 Table 7.1, an example of the decision tables used here, consists of the following objects: $U = \{x_1, \ldots, x_6\}$, $C = \{c_1, \ldots, c_6\}$, $D = \{d\}$, $V = \{0, 1, 2, 3\}$, and function $\rho : U \times (C \cup D) \to V$ describes values of objects for attributes such that $\rho(x1, c1) = 1$. Set D provides three decision classes: $D_1 = \{x_1, x_2, x_5\}$, $D_2 = \{x_3, x_4\}$, and $D_3 = \{x_6\}$.

7.2.2 Relative Reduct

Checking values of all condition attributes enable us to classify all discernible objects of the given decision table into corresponding decision classes. Not all condition attributes need be checked, however, because only certain condition attributes are required for classification and others are redundant. Minimal sets of condition attributes classifying all discernible objects into correct decision classes are called *relative reducts* of the decision table.

To introduce relative reducts, for any subset $X \subseteq C$ of condition attributes in decision table DT, we let:

$$POS_X(\mathcal{D}) = \bigcup_{D_i \in \mathcal{D}} \underline{X}(D_i). \tag{7.5}$$

Set $POS_X(\mathcal{D})$ is called the positive region of $\mathcal{D}$ by X. All objects $x \in POS_X(\mathcal{D})$ are classified into correct decision classes by checking all attributes in X. Specifically, set $POS_C(\mathcal{D})$ is the set of all discernible objects in DT.

Formally, set $A \subseteq C$ is called a relative reduct of decision table DT if set A satisfies the following conditions:

1. $POS_A(\mathcal{D}) = POS_C(\mathcal{D})$.
2. $POS_B(\mathcal{D}) \neq POS_C(\mathcal{D})$ for any proper subset $B \subset A$.

Note that decision tables generally have plural relative reducts and the part common to all relative reducts is called the core of the decision table.

Take, for example, the three relative reducts in Table 7.1 $\{c_3, c_5\}$, $\{c_5, c_6\}$, and $\{c_2, c_4, c_5\}$. Condition attribute c_5 appears in all relative reducts in Table 7.1, so the core of Table 7.1 is $\{c_5\}$.

7.2.3 Discernibility Matrix

The DM is used to compute all relative reducts in the decision table. Let DT be a decision table with $|U|$ objects, where $|U|$ is the cardinality of U. DM of DT is a symmetric $|U| \times |U|$ matrix whose element at the i-th row and j-th column is the following set of condition attributes for discerning between two objects x_i and x_j:

$$\delta_{ij} = \begin{cases} \{a \in C \mid \rho(x_i, a) \neq \rho(x_j, a)\}, \\ \exists d \in D \text{ s.t. } \rho(x_i, d) \neq \rho(x_j, d), \\ \text{and } \{x_i, x_j\} \cap POS_C(\mathcal{D}) \neq \emptyset. \\ \emptyset, \qquad\qquad\qquad\qquad \text{otherwise.} \end{cases} \tag{7.6}$$

Each element $a \in \delta_{ij}$ denotes that x_i and x_j are discernible by checking the value of a. Using the DM, we get all relative reducts of the decision table as follows:

1. Construct logical formula $L(\delta_{ij})$ from each nonempty set $\delta_{ij} = \{a_{k1}, \ldots, a_{kl}\}$ ($i > j$ and $l \geq 1$) in the DM:

$$L(\delta_{ij}) : a_{k1} \vee \cdots \vee a_{kl}. \tag{7.7}$$

2. Construct conjunctive normal form $\bigwedge_{i>j} L(\delta_{ij})$.
3. Transform the conjunctive normal form to the minimal disjunctive normal form:

$$\bigwedge_{i>j} L(\delta_{ij}) \equiv \bigvee_{p=1}^{s} \bigwedge_{q=1}^{t_p} a_{pq}. \tag{7.8}$$

4. For each conjunction $a_{p1} \wedge \cdots \wedge a_{pt_p}$ ($1 \leq p \leq s$) in the minimal disjunctive normal form, construct relative reduct $\{a_{p1}, \ldots, a_{pt_p}\}$.

Example 7.2 Table 7.2 describes the DM of the decision table in Table 7.1. Each nonempty set appearing in the matrix represents the set of condition attributes we check to discern corresponding objects. Set $\delta_{65} = \{c_3, c_5\}$, for example, shows that

Table 7.2 Discernibility matrix of Table 7.1

	x_1	x_2	x_3	x_4	x_5	x_6
x_1	$\emptyset$					
x_2	$\emptyset$	$\emptyset$				
x_3	$\{c_1, c_2, c_3, c_6\}$	$\{c_2, c_3, c_6\}$	$\emptyset$			
x_4	$\{c_1, c_2, c_3, c_4, c_5, c_6\}$	$\{c_3, c_4, c_5, c_6\}$	$\emptyset$	$\emptyset$		
x_5	$\emptyset$	$\emptyset$	$\{c_2, c_3, c_6\}$	$\{c_3, c_4, c_5, c_6\}$	$\emptyset$	
x_6	$\{c_1, c_2, c_5\}$	$\{c_5\}$	$\{c_2, c_3, c_5, c_6\}$	$\{c_3, c_4, c_6\}$	$\{c_3, c_5\}$	$\emptyset$

we distinguish between objects x_6 and x_5 by comparing the values of these objects at condition attribute c_3 or c_5.

Note that we omit upper triangular components of the DM in Table 7.2, because the DM is symmetric by definition. We construct a conjunctive normal form by connecting logical formulas based on nonempty elements in Table 7.2 by (7.7) and (7.8), and transform the conjunctive normal form to the minimal disjunctive normal form as follows:

$$
(c_1 \vee c_2 \vee c_3 \vee c_6) \wedge (c_2 \vee c_3 \vee c_6) \wedge \cdots \wedge (c_3 \vee c_5)
$$
$$
\equiv (c_3 \wedge c_5) \vee (c_5 \wedge c_6) \vee (c_2 \wedge c_4 \wedge c_5).
$$

This minimal disjunctive normal form yields the three relative reducts $\{c_3, c_5\}$, $\{c_5, c_6\}$, and $\{c_2, c_4, c_5\}$.

7.3 Evaluation of Relative Reducts Based on Classification Ability

To introduce the evaluation of relative reducts based on the classification ability of condition attributes, we focus on condition attributes appearing in relative reducts. We consider condition attributes having the following two properties for high classification ability:

- Condition attributes discerning objects of different decision classes as long as possible.
- Condition attributes not discerning objects of the same decision class as long as possible.

With this evaluation policy, for each condition attribute $a \in C$, we consider the following two sets:

$$
Dis(a) = \left\{ (x_i, x_j) \in U \times U \left| \begin{array}{l} \{x_i, x_j\} \cap Pos_C(\mathcal{D}) \neq \emptyset, \\ \rho(x_i, a) \neq \rho(x_j, a), \\ \rho(x_i, d) \neq \rho(x_j, d), \\ \exists d \in D, i > j \end{array} \right. \right\}, \tag{7.9}
$$

$$Indis(a) = \left\{ (x_i, x_j) \in U \times U \left| \begin{array}{l} \{x_i, x_j\} \cap Pos_C(\mathcal{D}) \neq \emptyset, \\ \rho(x_i, a) = \rho(x_j, a), \\ \rho(x_i, d) = \rho(x_j, d), \\ \forall d \in D, i > j \end{array} \right. \right\}. \qquad (7.10)$$

Set $Dis(a)$ describes a set of ordered pairs (x_i, x_j) such that x_i and x_j belong to different decision classes and are discerned by attribute a values. Set $Indis(a)$ describes a set of (x_i, x_j) such that x_i and x_j belong to the same decision class and are not discerned by a values. The definition of $Dis(a)$ and $Indis(a)$ clarifies that $(x_i, x_j) \in Dis(a)$ implies $(x_i, x_j) \notin Indis(a)$ or, equivalently, that $(x_i, x_j) \in Indis(a)$ implies $(x_i, x_j) \notin Dis(a)$.

For each condition attribute $a \in C$, we introduce the following criterion for classification ability:

$$Eval(a) \overset{\text{def}}{=} \begin{cases} |Dis(a)| + |Indis(a)| & \text{for } Dis(a) \neq \emptyset, \\ 0 & \text{for } Dis(a) = \emptyset. \end{cases} \qquad (7.11)$$

We regard that the higher the evaluation score $Eval(a)$, the higher the classification ability of condition attribute a.

$|Dis(a)|$ is easily confirmed to be the number of attribute a occurrences in the lower triangular part of the DM. We thus calculate value $|Dis(a)|$ when we construct the DM. We also calculate $|Indis(a)|$ simultaneously, enabling us to evaluate all condition attributes during DM construction. Note that we define $Eval(a) = 0$ if $Dis(a) = \emptyset$ holds, because condition attributes such as a do not work in classifying objects.

Proposition 7.1 *Let $DT = (U, C \cup D, V, \rho)$ be a decision table. For every condition attribute $a \in C$, the range of evaluation score $Eval(a)$ defined by (7.11) is:*

$$0 \leq Eval(a) \leq \frac{|U|(|U| - 1)}{2}. \qquad (7.12)$$

Proof Former part $0 \leq Eval(a)$ holds by the definition of $Eval(a)$. For the latter part, the maximum score is the case in which for every pair (x_i, x_j) of objects $(i > j)$, either $(x_i, x_j) \in Dis(a)$ or $(x_i, x_j) \in Indis(a)$ holds. The maximum score is number of pairs (x_i, x_j) with $1 \leq j < i \leq |U|$, so the latter part holds.

Let $B \subseteq C$ be a relative reduct of the given decision table. Evaluation score $Eval(B)$ of the relative reduct is defined by:

$$Eval(B) \overset{\text{def}}{=} \frac{1}{|B|} \sum_{a \in B} Eval(a), \qquad (7.13)$$

that is, the average of evaluation score of condition attributes in B. Similar to condition attributes, we assume that the higher the evaluation value $Eval(B)$, the higher the classification ability of relative reduct B.

Proposition 7.2 *For every relative reduct B of the given decision table, the range of evaluation score $Eval(B)$ defined by (7.13) is:*

$$0 < Eval(B) \leq \frac{|U|(|U|-1)}{2}. \tag{7.14}$$

Specifically, if $Eval(B) = |U|(|U|-1)/2$ holds for relative reduct B, the decision table is consistent and B is a singleton.

Proof For every relative reduct B, $Eval(a) > 0$ holds for all $a \in B$ because, by the definition of $Eval(a)$, no condition attribute a with $Eval(a) = 0$ appears in the DM, so such attributes do not belong to any relative reduct. This concludes former part $0 < Eval(B)$. The latter part is trivial from Proposition 7.1.

Next, assume relative reduct B with $Eval(B) = |U|(|U|-1)/2$. $Eval(a) = |U|(|U|-1)/2$ then holds for every $a \in B$, implying that for every pair (x_i, x_j) of objects $(i > j)$, $(x_i, x_j) \in Dis(a)$ holds if x_i and x_j belong to different decision classes. For any x_i and x_j in different decision classes, $\delta_{ij} \neq \emptyset$ and $a \in \delta_{ij}$ then hold and the decision table must be consistent and singleton $\{a\}$ must be a relative reduct, concluding $B = \{a\}$.

The converse of Proposition 7.2 is generally not satisfied, i.e., even though the decision table is consistent and relative reduct B a singleton, we may have $Eval(B) < |U|(|U|-1)/2$.

Yamaguchi [20] proposed improving Pawlak's attribute dependency [12] based on a similar policy used in evaluation criteria (7.11) and (7.13), but his formulation differs from this evaluation and is not used to evaluate relative reducts.

Example 7.3 Considering the evaluation of condition attributes and relative reducts in Table 7.1, we first construct for each condition attribute $c_i \in \{c_1, \ldots, c_6\}$, two sets $Dis(c_i)$ and $Indis(c_i)$ defined by (7.9) and (7.10). For condition attribute c_1, these sets are:

$$Dis(c1) = \{(x_3, x_1), (x_4, x_1), (x_6, x_1)\},$$
$$Indis(c1) = \{(x_5, x_2), (x_4, x_3)\}.$$

c_1 thus discerns between x_3 and x_1, x_4 and x_1, and x_6 and x_1, which we must discern, and does not discern x_5 and x_2 and x_4 and x_3, which need not be discerned. To evaluate the classification ability of c_1, we calculate the following evaluation score of c_1 by criterion (7.11):

$$Eval(c_1) = |Dis(c_1)| + |Indis(c_1)| = 3 + 2 = 5.$$

Similarly, other condition attributes have the following scores:

$$Eval(c_2) = 7, \quad Eval(c_3) = 11, \; Eval(c_4) = 7,$$
$$Eval(c_5) = 10, \; Eval(c_6) = 12.$$

We then consider that condition attribute c_6 has the highest classification ability relative to decision classes. c_6 actually discerns objects of decision class $D_2 = \{x_3, x_4\}$ from objects in $D_1 = \{x_1, x_2, x_5\}$ and $D_3 = \{x_6\}$ and c_6 does not discern any objects in D_1 or D_2.

Using condition attribute evaluation scores, we then evaluate three relative reducts of Table 7.1 using criterion (7.13):

$$Eval(\{c_3, c_5\}) = \frac{1}{2}(11 + 10) = 10.5,$$

$$Eval(\{c_5, c_6\}) = \frac{1}{2}(10 + 12) = 11,$$

$$Eval(\{c_2, c_4, c_5\}) = \frac{1}{3}(7 + 7 + 10) = 8.$$

Relative reduct $\{c_5, c_6\}$ has the highest score, so we consider it the best relative reduct.

7.4 Heuristic Algorithm for Computing a Relative Reduct Candidate

In this section, we propose a heuristic algorithm for computing a relative reduct candidate with a high evaluation score using the proposed relative reduct evaluation criterion (7.13).

For generating a relative reduct candidate $S \subseteq C$, we revise the current candidate S by adding a condition attribute with the currently highest evaluation score and update condition attribute evaluation scores until the candidate has nonempty intersections with all nonempty elements in the DM.

As stated in Sect. 7.3, value $|Dis(a)|$ used in (7.11) is the number of attribute $a \in C$ occurrences in the lower triangular part of the DM, so we update value $|Dis(a)|$ of each condition attribute a by revising the DM. To update $|Indis(a)|$ of each a, we introduce equivalent matrix (EM) of a given decision table. Formally, an EM of given decision table $DT = (U, C \cup D, V, \rho)$ is a $|U| \times |U|$ matrix whose element at the i-th row and j-th column is the following set ϵ_{ij} of condition attributes:

$$\epsilon_{ij} \overset{\text{def}}{=} \begin{cases} \{a \in C \mid \rho(x_i, a) = \rho(x_j, a)\}, \\ \quad \{x_i, x_j\} \cap Pos_C(\mathcal{D}) \neq \emptyset, \text{ and} \\ \quad \forall d \in D, \; \rho(x_i, d) = \rho(x_j, d). \\ \emptyset, \qquad\qquad\qquad \text{otherwise.} \end{cases} \tag{7.15}$$

Nonempty element ϵ_{ij} is the set of condition attributes wherein x_i and x_j belonging to the same decision class have the same values. The definition of EM easily confirms that $\epsilon_{ij} = \epsilon_{ji}$ and $\epsilon_{ii} = C$ for any $i, j \in \{1, \ldots, |U|\}$. Similar to the DM, the EM is symmetric and constructing the lower triangular part of the EM is sufficient for actual computation.

Using DM and EM, $Dis(a)$ and $Indis(a)$ definitions are revised as follows:

$$Dis(a) = \left\{ (x_i, x_j) \in U \times U \mid a \in \delta_{ij}, i > j \right\}, \tag{7.16}$$

$$Indis(a) = \left\{ (x_i, x_j) \in U \times U \mid a \in \epsilon_{ij}, i > j \right\}. \tag{7.17}$$

We calculate evaluation score $Eval(a)$ for each condition attribute $a \in C$, based on DM and EM. Note that Proposition 7.1 indicates that the sum of the number of elements in the lower triangular parts of the DM and EM is at most $|U|(|U| - 1)/2$.

We propose the following heuristic algorithm for computing a relative reduct candidate with a high evaluation score.

Algorithm for Computing a Relative Reduct Candidate
Input: Decision table $DT = (U, C \cup D, V, \rho)$.
Output: Relative reduct candidate $S \subseteq C$.

1. Compute discernibility matrix DM and equivalent matrix EM from DT.
2. $S = \bigcup \{\delta \in DM \mid |\delta| = 1\}$ and $C := C - S$.
3. If $S \cap \delta_{ij} \neq \emptyset$ holds for all nonempty elements $\delta_{ij} \in DM$ $(i > j)$, output S and quit.
4. For each nonempty $\delta_{ij} \in DM$ $(i > j)$, if $S \cap \delta_{ij} \neq \emptyset$ holds, then $\delta_{ij} := \emptyset$. Similarly, for each nonempty $\epsilon_{ij} \in EM$ $(i > j)$, if $S \cap \epsilon_{ij} \neq \emptyset$ holds, then $\epsilon_{ij} := \emptyset$.
5. For each $c \in C$, calculate evaluation score $Eval(c) = |Dis(c)| + |Indis(c)|$ using DM and EM.
6. Select condition attribute $c_h \in C$ with the highest evaluation score and $S := S \cup \{c_h\}$.
7. $C := C - \{c_h\}$ and return to Step 3.

Note that, for any sets X and Y, notation $X := Y$ means that we use set Y to replace set X.

In this algorithm, after constructing a DM and an EM from the given decision table, we construct the core of the decision table in Step 2 using the following property: condition attribute $a \in C$ is in the core if and only if element $\delta = \{a\}$ is in the DM [16]. In Step 3, we check whether the current candidate S discerns all discernible objects belonging to mutually different decision classes, and output S as a relative reduct candidate if S passes the check. In Steps 4 and 5, we revise the DM and EM for updating current evaluation scores of condition attributes. In Step 6, we revise the current candidate by adding the condition attribute with the currently highest evaluation score. For plural condition attributes with the currently highest evaluation score, we randomly select one attribute from currently best attributes. In Step 7, we remove the selected condition attribute from C and return to Step 3.

Table 7.3 Equivalent matrix of Table 7.1

	x_1	x_2	x_3	x_4	x_5	x_6
x_1	C					
x_2	$\{c_3, c_4, c_5, c_6\}$	C				
x_3	$\emptyset$	$\emptyset$	C			
x_4	$\emptyset$	$\emptyset$	$\{c_1, c_3, c_6\}$	C		
x_5	$\{c_4, c_5, c_6\}$	$\{c_1, c_2, c_4, c_5, c_6\}$	$\emptyset$	$\emptyset$	C	
x_6	$\emptyset$	$\emptyset$	$\emptyset$	$\emptyset$	$\emptyset$	C

Example 7.4 We compute a relative reduct candidate of Table 7.1 using the proposed algorithm.

We first compute a DM and an EM for Table 7.1 in Step 1. Table 7.3 represents the EM. Element $\epsilon_{43} = \{c_1, c_3, c_6\}$ means, for example, that values of objects x_4 and x_3 in the same decision class are mutually identical under the condition attributes c_1, c_3, and c_6. Note that the DM has already been presented in Table 7.2.

We then get core S of decision table in Step 2 as an initial relative reduct candidate. Only one singleton $\delta_{62} = \{c_5\}$ exists in DM, so we have core $S = \{c_5\}$, then revise set C of condition attributes to $C := \{c_1, c_2, c_3, c_4, c_6\}$.

In Step 3, we check whether current set $S = \{c_5\}$ is a relative reduct candidate. S does not pass the check, however, because $\delta_{31} \neq \emptyset$ and $\delta_{31} \cap S = \emptyset$ hold.

In Step 4, we replace the following elements in the DM by the empty set because these elements contain c_5:

$$\delta_{41} = C, \ \delta_{42} = \delta_{54} = \{c_3, c_4, c_5, c_6\},$$
$$\delta_{61} = \{c_1, c_2, c_5\}, \delta_{62} = \{c_5\}, \delta_{65} = \{c_3, c_5\}.$$

We also replace the following elements in EM by the empty set:

$$\epsilon_{21} = \{c_3, c_4, c_5, c_6\}, \ \epsilon_{51} = \{c_4, c_5, c_6\},$$
$$\epsilon_{52} = \{c_1, c_2, c_4, c_5, c_6\}.$$

These operations correspond to ignoring all elements in lower triangular DM parts in Table 7.2 and EM in Table 7.3 containing c_5.

In Step 5, we calculate current evaluation scores of condition attributes except for c_5 as follows:

$$Eval(c_1) = 2, \ Eval(c_2) = 3, \ Eval(c_3) = 5,$$
$$Eval(c_4) = 1, \ Eval(c_6) = 5.$$

In Step 6, we select the condition attribute with the highest scores, i.e., c_3 and c_6, so we must randomly select one of them. In this example, we select c_3 at random and update current set S to $S := \{c_3, c_5\}$. In Step 7, we remove selected attribute c_3 from C and return to Step 3.

Once the check of current set $S = \{c_3, c_5\}$ is passed in Step 3, we output S as a relative reduct candidate. Output $S = \{c_3, c_5\}$ is the second-best relative reduct in Table 7.1, indicating that this algorithm may not provide the best relative reduct via the criterion (7.13).

7.5 Experiments

We applied our heuristic algorithm to 13 UCI machine learning repository data sets [24]: audiology, hepatitis, lung-cancer, Monk1, Monk2, Monk3, soybean-s, soybean-l, SPECT-test, SPECT-train, tic-tac-toe, vote, and Zoo.

In describing experiment results, Table 7.4 lists dataset names in column 1, numbers of dataset condition attributes and objects in columns 2 and 3, numbers of all relative reducts of datasets in column 4, and output length such as numbers of attributes in output of our algorithm for each dataset in column 5. Parenthesized numbers in column 5 indicate that output includes redundant condition attributes, e.g., hepatitis dataset value (5) states that output contains some redundant attributes in five condition attributes. Column 6 shows the output evaluation score for the proposed criterion (7.13).

Note that symbol '-' of the hepatitis dataset in column 6 indicates that we did not evaluate hepatitis dataset output, because output was not a relative reduct. Column 7 shows output ranking among all relative reducts for each dataset using evaluation scores for the proposed criterion (7.13). Value "1" in column 7 of the Zoo dataset means, for example, that the output evaluation score for the proposed criterion is the highest score among the 33 relative reducts of the Zoo data. Symbol '-' in columns 4 and 7 shows that we could not compute all relative reducts of these datasets and could not rank output among all relative reducts using the proposed criterion.

We applied the QuickReduct algorithm [1] and Zhang et al.'s algorithm [22] to the 13 datasets in Table 7.4 to compare them to our heuristic algorithm, and compared the number of selected condition attributes and the score of outputs to those of the proposed algorithm. Table 7.5 gives output length and evaluation scores by QuickReduct (QR), Zhang et al.'s algorithm (ZWLHS), and our proposed algorithm (proposed) for each dataset.

QuickReduct uses random selection of condition attributes, so QuickReduct is applied 100 times to each dataset. Values without parentheses in column 2 show output length with the highest scores in the 100 repetitions described in column 5. Parenthesized values in column 2 show the length of the shortest output in the 100 repetitions, meaning that QuickReduct could not generate relative reducts of the dataset in this experiment. Note that, similar to the case in Table 7.4, symbol '-' in columns 5 and 7 means that we did not evaluate the score of outputs containing redundant condition attributes.

Table 7.4 Experiment results of the proposed algorithm

Data set	Attributes	Objects	Reducts	Length	Score	Ranking
Audiology	69	200	–	13	9080.31	–
Hepatitis	19	155	694	(5)	–	–
Lung-cancer	57	32	–	5	278.6	–
Monk1	6	124	1	3	4063	1
Monk2	6	169	1	6	6994.17	1
Monk3	6	122	1	4	4032.5	1
Soybean-s	35	47	–	2	867	–
Soybean-l	35	307	–	9	31081.45	–
SPECT-test	22	187	40	12	8971.34	20
SPECT-train	22	80	26	11	1668.18	5
Tic-tac-toe	9	958	9	8	227541.5	1
Vote	16	435	3	9	59649.78	1
Zoo	17	101	33	5	3542.8	1

Table 7.5 Comparison with other algorithms

Data set	Length			Score		
	QR	ZWLHS	Proposed	QR	ZWLHS	Proposed
Audiology	(25)	13	13	–	9080.31	9080.31
Hepatitis	6	3	(5)	6454	5507.67	–
Lung-cancer	5	4	5	272.8	288.25	278.6
Monk1	3	3	3	4063	4063	4063
Monk2	6	6	6	6994.17	6994.17	6994.17
Monk3	4	4	4	4032.5	4032.5	4032.5
Soybean-s	2	2	2	963	867	867
Soybean-l	(14)	9	9	–	31081.45	31081.45
SPECT-test	(14)	12	12	–	9077.17	8971.34
SPECT-train	(12)	11	11	–	1668.18	1668.18
Tic-tac-toe	(9)	8	8	–	227087.13	227541.5
Vote	13	9	9	56853.23	59649.78	59649.78
Zoo	6	5	5	3044.84	3542.8	3542.8

7.6　Discussion

As shown in Table 7.4, for Monk1, Monk2, Monk3, tic-tac-toe, vote, and Zoo, our algorithm generated relative reducts having the best evaluation for the proposed criterion (7.13). In other cases, except for hepatitis, our algorithm generated a relative reduct, i.e., no output of our proposal had any redundant condition attributes. The larger the set U of objects, clearly the higher the output score, because the maximum score depends on $|U|$, as shown in (7.14).

The case of the hepatitis dataset indicates that proposed algorithm output S is merely a relative reduct candidate that may not even be a relative reduct. Formally, S may not satisfy relative reduct condition 2, even though S is guaranteed to satisfy relative reduct condition 1 because the proposed algorithm does not check attribute redundancy in S for classification. We introduce redundancy checks ensuring that S is a relative reduct, but computational complexity of redundancy check is $O\left(2^{|S|}\right)$ in the worst case, and we omit redundancy checking to avoid high computational cost.

In the comparison of the proposed algorithm with QuickReduct and Zhang et al.'s algorithms, the proposed algorithm generated output having better evaluation than QuickReduct output in many datasets. Output evaluation scores for Zhang et al.'s algorithm were almost equal to output scores for our algorithm, probably because Zhang et al.'s algorithm also uses the number of condition attribute a occurrences in the DM, i.e., $|Dis(a)|$, as a main criterion for selecting condition attributes. The number of nonempty elements in the DM generally exceeds the number of nonempty elements in the EM, so value $|Dis(a)|$ dominates the proposed criterion (7.13), and proposed algorithm output resembles that of Zhang et al.'s algorithm.

Our algorithm is guaranteed to stop by computing relative reduct candidate S because our algorithm repeats Steps 3–7 $|C|$ times at most, and each step requires a finite step. Considering computational complexity of our algorithm, we check the computational cost of each step. Computational costs for constructing the DM and the EM, i.e., in Step 1, are both $O(|U|^2|C|)$ and the DM and EM are computed simultaneously, so Step 1 costs $O(|U|^2|C|)$. Because the number of elements in lower triangular parts of the DM and EM equal $|U|(|U|-1)/2$, at most, and attributes in each DM and EM element are at most $|C|$, the cost of computing the core in Step 2 is $O(|U|^2)$ and that of each step in Steps 3–5 is at most $O(|U|^2|C|)$. Steps 6 and 7 cost at most $O(|C|)$, so total computational complexity of the proposed algorithm is at most $O(|U|^2|C|^2)$.

7.7 Conclusion

We have proposed a relative reduct evaluation criterion based on the classification ability of condition attributes in relative reducts, together with a heuristic algorithm for computing a relative reduct candidate that results in good evaluation using the proposed evaluation criterion. The algorithm's computational complexity is at most $O(|U|^2|C|^2)$. Applying our algorithm to 13 UCI machine learning repository datasets, we generated output resulting in good evaluations for almost all datasets. Our algorithm was also compared to the QuickReduct and Zhang et al.'s algorithms. We now plan process refinement and more experiments using extensive datasets.

References

1. Chouchoulas, A., Shen, Q.: Rough set-aided keyword reduction for text categorization. Appl. Artif. Intell. **15**, 843–873 (2001)
2. Ge, H., Yang, C.: An efficient attribute reduction algorithm based on concentrated ordered discernibility set. In: Proceedings of 2008 International Conference on Computer Science and Software Engineering, pp. 332–335. IEEE (2008)
3. Guan, J.W., Bell, D.A.: Rough computational methods for information systems. Artif. Intell. **105**, 77–103 (1998)
4. Hedar, A.H., Wang, J., Fukushima, M.: Tabu search for attribute reduction in rough set theory. Soft Couping **12**, 909–918 (2008)
5. Hu, F., Wang, G., Feng, L.: Fast knowledge reduction algorithms based on quick sort. Rough Sets and Knowledge Technology. LNAI, vol. 5009, pp. 72–79. Springer, Berlin (2008)
6. Hu, K., Diao, L., Lu, Y., Shi, C.: A heuristic optiaml reduct algorithm. In: Proceedings of IDEAL 2000. LNSC, vol. 1983, pp. 139–144. Springer, Berlin (2000)
7. Kryszkiewicz, M., Lasek, P.: FUN: fast discovery of minimal sets of attributes gunctionally determining a decision attribute. Transactions on Rough Sets IX. LNCS, vol. 5390, pp. 76–95. Springer, Berlin (2008)
8. Kudo, Y., Murai, T.: A heuristic algorithm for selective calculation of a better relative reduct in rough set theory. In: Nakamatsu, K. et al. (eds.) New Advances in Intelligent Decision Technologies. SCI, vol. 199, pp. 555–564. Springer, Berlin (2009)
9. Kudo, Y., Murai, T.: A heuristic algorithm for attribute reduction based on discernibility and equivalence by attributes. In: Proceedings of the 6th International Conference on Modeling Decisions for Artificial Intelligence. LNAI, vol. 5861, pp. 351–359. Springer, Berlin (2009)
10. Mori, N., Tanaka, H., Inoue, K.: Rough sets and Kansei—knowledge acquisition and reasoning from Kansei data. Kaibundo (2004) (in Japanese)
11. Pawlak, Z.: Rough sets. Int. J. Comput. Inf. Sci. **11**, 341–356 (1982)
12. Pawlak, Z.: On rough dependency of attributes in information systems. Bull. Pol. Acad. Sci., Tech. Sci. **33**, 481–485 (1985)
13. Pawlak, Z.: Rough Sets: Theoretical Aspects of Reasoning about Data. Kluwer Academic Publishers, Dordrecht (1991)
14. Pawlak, Z., Słowiński, R.: Rough set approach to multi-attribute decision analysis. Eur. J. Oper. Res. **74**, 443–459 (1994)
15. Polkowski, L.: Rough Sets: Mathematical Foundations. Advances in Soft Computing. Physica-Verlag, Heidelberg (2002)
16. Skowron, A., Rauszer, C.M.: The discernibility matrix and functions in information systems. In: Słowiński, R. (ed.) Intelligent Decision Support: Handbook of Application and Advance of the Rough Set Theory, pp. 331–362. Kluwer, Dordrecht (1992)
17. Tan, S., Xu, H., Gu, J.: Efficient algorithms for attributes reduction problem. Int. J. Innov. Comput., Inf. Control. **1**, 767–777 (2005)
18. Xu, J., Sun, L.: New reduction algorithm based on decision power of decision table. Rough Sets and Knowledge Technology. LNAI, vol. 5009, pp. 180–188. Springer, Berlin (2008)
19. Xu, Z., Zhang, C., Zhang, S., Song, W., Yang, B.: Efficient attribute reduction based on discernibility matrix. Rough Sets and Knowledge Technology. LNAI, vol. 4481, pp. 13–21. Springer, Berlin (2007)
20. Yamaguchi, D.: Attribute dependency functions considering data efficiency. Int. J. Approx. Reason. **51**, 89–98 (2009)
21. Yao, Y.Y., Zhao, Y., Wang, J., Han, S.: A model of user-oriented reduct construction for machine learning. Transactions on Rough Sets VIII. LNCS, vol. 5084, pp. 332–351. Springer, Berlin (2008)
22. Zhang, J. Wang, J., Li, D., He, H., Sun, J.: A new heuristic reduct algorithm based on rough sets theory. In: Proceedings of WAIM 2003. LNCS, vol. 2762, pp. 247–253. Springer (2003)

23. Zhu, F., Zhang, Y., Zhao, L.: An efficient attribute reduction in decision information systems. In: Proceedings of 2008 International Conference on Computer Science and Software Engineering, pp. 466–469. IEEE (2008)
24. http://archive.ics.uci.edu/ml/index.html

Chapter 8
An Evaluation Method of Relative Reducts Based on Roughness of Partitions

Abstract This chapter, from the viewpoint of approximation, introduces an evaluation criterion for relative reducts using roughness of partitions constructed from them. The outline of relative reduct evaluation we propose is: "Good" relative reducts = relative reducts that provide partitions with approximations as rough and correct as possible. In this sense, we think that evaluation of relative reducts is strictly concerned with evaluation of roughness of approximation.

Keywords Rough set · Relative reduct · Decision rule

8.1 Introduction

In *rough set theory* introduced by Pawlak [8, 9], set-theoretical approximation of concepts and reasoning about data are the two main topics. In the former, lower and upper approximations of concepts and their evaluations are the main topics. Accuracy, quality of approximation, and quality of partition are well-known criteria in evaluation of approximations; these criteria are based on the correctness of the approximation. However, the roughness of the approximation is not explicitly treated in these criteria.

In reasoning about data, the relative reduct is one of the most important concepts for rule generation based on rough set theory, and many methods for exhaustive or heuristic calculation of relative reducts have been proposed [1, 4–7, 9, 11–13, 15–17]. As an evaluation criterion for relative reducts, the cardinality of a relative reduct, i.e., the number of attributes in the relative reduct, is typical and is widely used (for example, in [5–7, 15, 17]). In addition, other kinds of criteria related to evaluation of partitions are also considered with respect to the following evaluation functions: a normalized decision function generated from a relative reduct B [11], the information entropy $H(B)$ of B [12], and the number of decision rules induced from B [14].

In this chapter, we consider evaluating relative reducts based on the roughness of partitions constructed from them. The outline of relative reduct evaluation we propose is: "Good" relative reducts = relative reducts that provide partitions with approxi-

© Springer Nature Switzerland AG 2020
S. Akama et al., *Topics in Rough Set Theory*, Intelligent Systems
Reference Library 168, https://doi.org/10.1007/978-3-030-29566-0_8

mations as rough and correct as possible. In this sense, we think that evaluation of relative reducts is strictly concerned with evaluation of roughness of approximation.

The rest of the chapter is structured as follows. In Sect. 8.2, we review the foundations of rough set theory as background for this chapter. In Sect. 8.3, we derive some properties related to roughness of partition and the average coverage of decision rules, and propose an evaluation criterion of relative reducts based on roughness of partition. In Sect. 8.4, we demonstrate the proposed method for evaluating relative reducts. Finally, we discuss the results of this paper and present our conclusions in Sect. 8.5.

8.2 Rough Set

We review the foundations of rough set theory as background for this chapter. The contents of this section are based on [10].

In rough set data analysis, objects as targets of analysis are illustrated by a combination of multiple attributes and their values and is represented by the following *decision table* (U, C, d), where U is the set of objects, C is the set of condition attributes such that each attribute $a \in C$ is a function $a : U \to V_a$ from U to the value set V_a of a, and d is a function $d : U \to V_d$ called the decision attribute.

The *indiscernibility relation* R_B on U with respect to a subset $B \subseteq C$ is defined by

$$(x, y) \in R_B \iff a(x) = a(y), \forall a \in B. \tag{8.1}$$

It is easy to confirm that the indiscernibility relation R_B is an equivalence relation on U. The equivalence class $[x]_B$ of $x \in U$ by R_B is the set of objects which are not discernible with even though they use all attributes in B.

Any indiscernibility relation provides a partition of U. We denote the quotient set of U, i.e., a partition of U, with respect to an equivalence relation R by U/R. In particular, the partition $\mathcal{D} = \{D_1, ..., D_n\}$ provided by the indiscernibility relation $R_{\{d\}}$ with respect to the decision attribute is called the set of decision classes.

For any decision class $D_i (1 \leq i \leq m)$, the *lower approximation* $\underline{B}(D_i)$ and the *upper approximation* $\overline{B}(D_i)$ of D_i with respect to the indiscernibility relation R_B are defined as follows, respectively:

$$\underline{B}(D_i) = \{x \in U \mid [x]_B \subseteq D_i\}, \tag{8.2}$$
$$\overline{B}(D_i) = \{x \in U \mid [x]_B \cap D_i \neq \emptyset\}. \tag{8.3}$$

A pair $(\underline{B}(D_i), \overline{B}(D_i))$ is called a *rough set* of D_i with respect to R_B.

As evaluation criteria of approximation, the accuracy measure and the quality of approximation are well known. Moreover, the quality of partition by R_B with respect to the set of decision classes $\mathcal{D} = \{D_1, ..., D_m\}$ is also defined as

Table 8.1 Decision table

	$c1$	$c2$	$c3$	$c4$	$c5$	$c6$	d
$x1$	1	1	1	1	1	2	1
$x2$	2	2	1	1	1	2	1
$x3$	2	3	2	1	2	1	2
$x4$	2	2	2	2	2	1	2
$x5$	2	2	3	1	1	2	1
$x6$	1	2	1	1	2	2	3

$$\gamma_B(\mathcal{D}) = \frac{\sum_{i=1}^{m} |\underline{B}(D_i)|}{|U|}, \tag{8.4}$$

where $|X|$ is the cardinality of the set X.

Table 8.1 represents a decision table consisting of the set of objects $U = \{x_1, ..., x_6\}$, the set of condition attributes $C = \{c_1, ..., c_6\}$, and the decision attribute d.

For example, the attribute $c2$ is the function $c2 : U \rightarrow \{1, 2, 3\}$, and the value of an object $x3 \in U$ at $c2$ is 3, i.e., $c2(x3) = 3$. Moreover, the decision attribute d provides the following three decision classes: $D_1 = \{x1, x2, x5\}$, $D_2 = \{x3, x4\}$, and $D_3 = \{x6\}$.

In this chapter, we denote a decision rule constructed from a subset $B \subseteq C$ of condition attributes, the decision attribute d, and an object $x \in U$ by

$$(B, x) \rightarrow (d, x).$$

The concepts of cetainty and coverage are well-known criteria in evaluation of decision rules. For any decision rule $(B, x) \rightarrow (d, x)$, the score of certainty $Cer(\cdot)$ and the score of coverage $Cov(\cdot)$ are defined as follows:

$$Cer((B, x) \rightarrow (d, x)) = \frac{|[x]_B \cap D_i|}{|[x]_B|}, \tag{8.5}$$

$$Cov((B, x) \rightarrow (d, x)) = \frac{|[x]_B \cap D_i|}{|D_i|}. \tag{8.6}$$

where the set D_i is the decision class such that $x \in D_i$.

For example, a decision rule $(B, x) \rightarrow (d, x)$ constructed from a set $B = \{c2, c3\}$, the decision attribute d, and an object $x1 \in U$ has the form

$$(c2 = 1) \wedge (c3 = 1) \rightarrow (d = 1)$$

and its certainty is 1 and the coverage is 1/3.

Relative reducts are minimal subsets of condition attributes that provide the same result of classification by the indiscernibility relation R_C with respect to the set C of all condition attributes. Formally, a relative reduct for the partition $\mathcal{D}$ is a set of condition attributes $A \subseteq C$ that satisfies the following two conditions:

1. $\text{POS}_A(\mathcal{D}) = \text{POS}_C(\mathcal{D})$.
2. $\text{POS}_B(\mathcal{D}) \neq \text{POS}_C(\mathcal{D})$ for any proper subset $B \subset A$,

where $POS_X(\mathcal{D}) = \bigcup_{D_i \in \mathcal{D}} \underline{B}(D_i)$ is the positive region of $\mathcal{D}$ by $X \subseteq C$.

For example, there are the following seven relative reducts of Table 8.1: $\{c1, c3\}$, $\{c3, c5\}$, $\{c5, c6\}$, $\{c1, c2, c3\}$, $\{c1, c2, c4\}$, $\{c1, c2, c6\}$, and $\{c2, c4, c5\}$.

8.3   Evaluation of Relative Reducts Using Partitions

In this section, we propose an evaluation method for relative reducts using partitions constructed from them. To summarize relative reduct evaluation, we consider that rougher partitions constructed by a relative reduct lead to better evaluation of the relative reduct.

However, the quality of partition defined by (8.4) does not consider roughness of partition, although it considers correctness of approximation. In fact, all the relative reducts of a consistent decision table, i.e., a decision table that satisfies $\text{POS}_C(\mathcal{D}) = U$, provide crisp approximations of all decision classes, i.e., $\underline{A}(D_i) = D_i = \overline{A}(D_i)$ for any relative reduct A and any decision class D_i, even though the difference of roughness of partitions based on relative reducts.

For example, because the decision table illustrated by Table 8.1 is consistent, all the relative reducts of Table 8.1 provide crisp approximation, and it is easy to confirm that the score of quality of partition by any relative reduct in Table 8.1 is equal to 1. However, the roughness of partitions based on the relative reducts differs, as follows:

- Partitions by the relative reduct $\{c1, c5\}$: $\{\{x1\}, \{x2, x5\}, \{x3, x4\}, \{x6\}\}$.
- Partitions by the relative reduct $\{c3, c5\}$: $\{\{x1, x2\}, \{x3, x4\}, \{x5\}, \{x6\}\}$.
- Partitions by the relative reduct $\{c5, c6\}$: $\{\{x1, x2, x5\}, \{x3, x4\}, \{x6\}\}$.
- Partitions by the relative reduct $\{c1, c2, c3\}$: $\{\{x1\}, \{x2\}, \{x3\}, \{x4\}, \{x5\}, \{x6\}\}$.
- Partitions by the relative reduct $\{c1, c2, c4\}$: $\{\{x1\}, \{x2, x5\}, \{x3\}, \{x4\}, \{x6\}\}$.
- Partitions by the relative reduct $\{c1, c2, c6\}$: $\{\{x1\}, \{x2, x5\}, \{x3\}, \{x4\}, \{x6\}\}$.
- Partitions by the relative reduct $\{c2, c4, c5\}$: $\{\{x1\}, \{x2, x5\}, \{x3\}, \{x4\}, \{x6\}\}$.

In particular, all equivalence classes in the partition by $\{c1, c2, c3\}$ are singletons. Thus, the quality of approximation is not suitable for evaluating roughness of partition.

In addition, from the viewpoint of rule generation, such rough partitions constructed from relative reducts provide decision rules with higher scores of coverage than those of coverage of decision rules based on fine partitions. Moreover, the correctness of partitions based on relative reducts is guaranteed, because each relative reduct provides the same approximation as the one based on the set of all condition attributes. Thus, we consider evaluating relative reducts using the coverage of decision rules constructed from them.

Here, we consider deriving some relationship between the roughness of partitions based on relative reducts and the coverage of decision rules based on them. Suppose we fix a non-empty subset of condition attributes $B \subseteq C$. For any equivalence class $[x]_B \in U/R_B$, we define the set

$$Dec_B([x]_B) \stackrel{\text{def}}{=} \{D_i \in \mathcal{D} \mid [x]_B \cap D_i \neq \emptyset\}. \tag{8.7}$$

The set $Dec_B([x]_B)$ corresponds to the set of conclusions of decision rules, with the formula (B, x) as the antecedent. Thus, the value defined as

$$N_B \stackrel{\text{def}}{=} \sum_{[x]_B \in U/R_B} |Dec_B([x]_B)|. \tag{8.8}$$

is the sum of the number of all decision rules constructed from B. Similarly, for any decision class $D_i \in \mathcal{D}$, we define the set

$$Cond_B(D_i) \stackrel{\text{def}}{=} \{[x]_B \in U/R_B \mid [x]_B \cap D_i \neq \emptyset\}. \tag{8.9}$$

The set $Cond_B(D_i)$ corresponds to the set of antecedents of decision rules with the formula (d, y) for some $y \in D_i$ as the conclusion. From the definitions of sets $Dec_B([x]_B)$ and $Cond_B(D_i)$, the following relationship is obvious:

$$D_i \in Dec_B([x]_B) \Leftrightarrow [x]_B \in Cond_B(D_i). \tag{8.10}$$

For evaluation of relative reducts using criteria involving decision rules, we use the following properties of the certainty and coverage of decision rules.

Lemma 8.1 *Let $B \subseteq C$ be a non-empty subset of condition attributes. For any equivalence class $[x]_B \in U/R_B$, the sum of the certainty of all decision rules constructed from the equivalence class $[x]_B$ and decision classes $D_i \in Dec_B([x]_B)$ is equal to 1; i.e., the following equation is satisfied:*

$$\sum_{D_i \in Dec_B([x]_B)} \frac{|[x]_B \cap D_i|}{|[x]_B|} = 1. \tag{8.11}$$

Similarly, for any decision class $D_i \in \mathcal{D}$, the sum of the coverage of all decision rules constructed from equivalence classes $[x]_B \in Cond_B(D_i)$ and the decision class D_i is equal to 1; i.e., the following equation is satisfied:

$$\sum_{[x]_B \in Cond_B(D_i)} \frac{|[x]_B \cap D_i|}{|D_i|} = 1. \tag{8.12}$$

Proof Here, we derive equation (8.11). Because the set $\mathcal{D} = \{D_1, ..., D_m\}$ of decision classes is a partition on U, it is clear that the set of non-empty intersections $[x]_B \cap D_i$ of decision classes D_i and a given equivalence class $[x]_B \in U/R_B$ is a partition on the equivalence class $[x]_B$. Thus, we have the following equation from (8.7):

$$[x]_B = \bigcup_{D_i \in Dec_B([x]_B)} ([x]_B \cap D_i) \tag{8.13}$$

Moreover, any two intersections $[x]_B \cap D_i$ and $[x]_B \cap D_j$ are disjoint from each other if the decision classes are different; therefore, we can simplify the left side of equation (8.11) as follows:

$$\sum_{D_i \in Dec_B([x]_B)} \frac{|[x]_B \cap D_i|}{|[x]_B|} = \frac{|[x]_B \cap D_{j1}| + \cdots + |[x]_B \cap D_{ji}|}{|[x]_B|} \quad (\forall D_{jk} \in Dec_B([x]_B))$$

$$= \frac{|([x]_B \cap D_{j1}) \cup \cdots \cup ([x]_B \cap D_{ji})|}{|[x]_B|}$$

$$= \frac{|[x]_B|}{|[x]_B|} \quad (by \ (8.13))$$

$$= 1.$$

This concludes the derivation of (8.11).

Equation (8.12) being derived in a similar way, we omit the proof.

Lemma 8.1 exhibits useful properties for considering the average certainty and the average coverage of decision rules constructed from non-empty subsets of condition attributes. This is because Lemma 8.1 indicates that the sum of the certainty (coverage) of all decision rules constructed from a given equivalence class (a fixed decision class) and decision classes (equivalence classes) relevant to the equivalence class (the decision class) is equal to 1, even though the detailed scores of certainty and coverage may differ because different equivalence classes and decision classes are used. Therefore, we do not need to consider detailed scores of certainty and coverage of each decision rule to calculate the average certainty and the average coverage.

We show the following theorem about the average certainty and the average coverage of all decision rules constructed from any non-empty subset of condition attributes. This theorem provides a basis for evaluating relative reducts we consider.

Theorem 8.1 *For any non-empty subset $B \subseteq C$ of condition attributes, the average of the certainty $ACer(B)$ of all decision rules $(B, x) \rightarrow (d, x)$ $(\forall x \in U)$ constructed from equivalence classes $[x]_B \in U/R_B$ and decision classes $[x]_d \in \mathcal{D}$ is calculated using*

$$ACer(B) = \frac{|U/R_B|}{N_B}, \tag{8.14}$$

where N_B is the number of all decision rules constructed from B defined by (8.8).

Similarly, the average of the coverage is calculated as

$$ACov(B) = \frac{|\mathcal{D}|}{N_B} \tag{8.15}$$

Proof Here, we derive equation (8.15). Because $\mathcal{D} = \{D_1., ..., D_m\}$ is a partition on U, we can consider the sum of the coverage of all decision rules constructed from B by treating all decision classes in (8.12). Thus, using the number N_B of all decision rules defined by (8.8), we simplify the average coverage of all decision rules constructed from B as

$$ACov(B) = \frac{1}{N_B} \sum_{D_i \in \mathcal{D}} \left(\sum_{[x]_B \in Cond_B(D_i)} \frac{|[x]_B \cap D_i|}{|D_i|} \right) \tag{8.16}$$

$$= \frac{1}{N_B} \sum_{D_i \in \mathcal{D}} 1 \quad (by\ (8.12))$$

$$= \frac{|\mathcal{D}|}{N_B}. \tag{8.17}$$

This concludes the derivation of (8.15).

Equation (8.14) being derived in a similar way, we omit the proof.

Theorem 8.1 demonstrates that we can calculate the average certainty and the average coverage of decision rules based on any non-empty set $B \subseteq C$ by the following three parameters: the number of equivalence classes based on B, the number of decision classes, and the number of decision rules constructed from B. In particular, because the number of decision classes is uniquely determined for any decision table, the average coverage depends only on the number of decision rules.

By Theorem 8.1, if we use a relative reduct as a subset of condition attributes in any consistent decision table, the average certainty value of decision rules constructed from the relative reduct is equal to 1. In addition, the smaller the number of decision rules constructed from the relative reduct, the higher the average coverage of decision rules.

Moreover, from Theorem 8.2, we can derive the relationship between the roughness of partitions based on relative reducts and the coverage of decision rules based on relative reducts. We use the following lemma to derive the relationship.

Lemma 8.2 *Let $B \subseteq C$ be a non-empty set of condition attributes, and D_i be a decision class. For any object $x \in U$, we have $x \in \underline{B}(D_i)$ iff $Dec_B([x]_B) = \{D_i\}$.*

Proof Suppose $x \in \underline{B}(D_i)$. By the definition of lower approximation, we have $[x]_B \subseteq D_i$, which implies $[x]_B \cap D_i \neq \emptyset$ and $[x]_B \cap D_j = \emptyset$ for any $D_j \neq D_i$. This concludes $Dec_B([x]_B) = \{D_i\}$. Conversely, suppose $Dec_B([x]_B) = \{D_i\}$. This indicates that there is just one decision class D_i such that $[x]_B \cap D_i \neq \emptyset$, and $[x]_B \cap D_j = \emptyset$ for any other decision class $D_j \neq D_i$. This implies $[x]_B \subseteq D_i$, and thus we have $x \in \underline{B}(D_i)$.

Theorem 8.2 *Let E and F be relative reducts of a given decision table. The following properties are satisfied:*

1. *If $R_E \subset R_F$ holds, i.e., the partition U/R_F is rougher than the partition U/R_E, then $ACov(E) \leq ACov(F)$ holds.*
2. *If the given decision table is consistent and $R_E \subset R_F$ holds, then $ACov(E) < ACov(F)$ also holds.*

Proof 1. By Theorem 8.1, it is sufficient to show that $R_E \subset R_F$ implies $N_E \geq N_F$, as defined by (8.8). It is easy to confirm that the condition $R_E \subset R_F$ implies $[x]_E \subseteq [x]_F$ for any $x \in U$, and there is at least one object $y \in U$ such that $[y]_F = \bigcup_{j=1}^{p}[y_j]_E$ ($p \geq 2$). Because the set $Dec_F([y]_F)$ contains at least one decision class, for each decision class $D_i \in Dec_F([y]_F)$, we have $D_i \cap [y]_F = \bigcup_{j=1}^{p}(D_i \cap [y_j]_E)$. Then, the intersection $D_i \cap [y]_F$ is either identical to an intersection $D_i \cap [y_j]_E$ by just one equivalence class $[y_j]_E$ or the union of plural non-empty intersections $D_i \cap [y_j]_E$. This implies the inequality $|Dec_F([y]_F)| \leq \sum_{j=1}^{p}|Dec_E([y_j]_E)|$, which concludes $N_E \geq N_F$.

2. Suppose the given decision table is consistent and we have $R_E \subset R_F$. By this condition, for any $y \in U$ with $[y]_F = \bigcup_{j=1}^{p}[y_j]_E$ ($p \geq 2$), $y \in D_i$ implies $y \in \underline{F}(D_i)$. By Lemma 8.2, we have $Dec_F([y]_F) = \{D_i\}$. Thus, similar to the case for 1, the following inequality $1 = |Dec_F([y]_F)| < \sum_{j=1}^{p}|Dec_E([y]_j]_E)|$ is satisfied. This concludes $N_E > N_F$, and therefore we have $ACov(E) < ACov(F)$.

Theorem 8.2 guarantees that relative reducts that provide rougher partitions receive better evaluations than those that provide finer partitions.

Combining Theorems 8.1 and 8.2, we can evaluate relative reducts of a given decision table by calculating the average coverage of decision rules constructed from the relative reducts. Therefore, we propose to use the average coverage of decision rules constructed from relative reducts as an evaluation criterion of the relative reducts based on roughness of partitions.

For example, we evaluate the relative reducts of Table 8.1 by calculating the average coverage of all decision rules constructed from these relative reducts. To calculate the average coverage, we need to get the number of decision rules constructed from each relative reduct. In the case of the relative reduct $A = \{c1, c5\}$, as we have illustrated in this section, the partition U/R_A consists of the following four equivalence classes: $\{x1\}$, $\{x2, x5\}$, $\{x3, x4\}$ and $\{x6\}$. For each equivalence class, the set of decision classes $Dec_A(\cdot)$ defined by (8.7) is calculated as

$$Dec_A([x1]_A) = \{D_1\}, Dec_A([x2]_A) = \{D_1\}, Dec_A([x3]_A) = \{D_2\}, Dec_A([x6]_A) = \{D_3\},$$

Thus, the number N_A of all decision rules is 4, and therefore, according to Theorem 8.1, the evaluation score of A is 3/4 because the number of decision classes in Table 8.1 is 3.

Table 8.2 Average certainty and average coverage calculated from relative reducts of Table 8.1

Relative reducts	Avg. of certainty	Avg. of coverage
$\{c1, c5\}$	1	0.75
$\{c3, c5\}$	1	0.75
$\{c5, c6\}$	1	1
$\{c1, c2, c3\}$	1	0.5
$\{c1, c2, c4\}$	1	0.6
$\{c1, c2, c6\}$	1	0.6
$\{c2, c4, c5\}$	1	0.6

We can construct the following four decision rules from $A = \{c1, c5\}$:

- $(c1 = 1) \wedge (c5 = 1) \rightarrow (d = 1)$, Certainty = 1, Coverage = 1/3.
- $(c1 = 2) \wedge (c5 = 1) \rightarrow (d = 1)$, Certainty = 1, Coverage = 2/3.
- $(c1 = 2) \wedge (c5 = 2) \rightarrow (d = 2)$, Certainty = 1, Coverage = 1.
- $(c1 = 1) \wedge (c5 = 2) \rightarrow (d = 3)$, Certainty = 1, Coverage = 1.

The average certainty of these rules is $(1 + 1 + 1 + 1)/4 = 1$, and this score is equal to the value "the number of equivalence classes/the number of decision rules" from Theorem 8.1. Moreover, the average coverage is $(1/3 + 2/3 + 1 + 1)/4 = 3/4$, and it is also equal to the value "the number of decision classes/the number of decision rules" from Theorem 8.1.

Table 8.2 shows the average scores of certainty and the average scores of coverage for each relative reduct. On the basis of this result, we regard the relative reduct $\{c5, c6\}$ to be the best one that provides the roughest and most correct approximations of decision classes. Actually, the partition constructed from the relative reduct $\{c5, c6\}$ is identical to the partition $U/R_{\{d\}}$, i.e., the set $\mathcal{D}$ of all decision classes.

8.4 Example

To demonstrate our method of evaluation of relative reducts, we apply the proposed method to a dataset, Zoo [2], in the UCI Machine Learning Repository (http://archive. ics.uci.edu/ml/). The Zoo data consists of 101 samples and 17 attributes with discrete values.

We use the attribute *type* as the decision attribute, and the remaining 16 attributes (*hair, feathers, eggs, milk, airborne, aquatic, predator, toothed, backbone, breathes, venomous, fins, legs, tail, domestic*, and *catsize*) as condition attributes. The decision attribute *type* provides seven decision classes corresponding to kinds of animals: mammal, bird, reptile, fish, amphibian, insect, and other invertebrates. Note that the decision table based on the Zoo data is consistent.

Consequently, there are 33 relative reducts in the Zoo data. These include 7 relative reducts that consist of five attributes, 18 that consist of six, and 8 that consist of seven.

Table 8.3 Experimental results of evaluation of relative reducts from the Zoo data

Number of attributes	Number of relative reducts	Maximum score of evaluation	Minimum score of evaluation	Average of evaluation
5	7	0.35	0.259	0.306
6	18	0.318	0.189	0.260
7	8	0.233	0.167	0.201
Total	33	0.35	0.167	0.260

Table 8.3 presents the experimental results of evaluation of relative reducts consisting of five, six, and seven attributes. The maximum, minimum, and average scores of evaluation of the 33 relative reducts are 0.35, 0.167, and 0.260, respectively.

These results indicate that the smaller the number of attributes in a relative reduct, the higher the score of evaluation in general. However, even when the number of attributes is identical, big differences occur between the scores of evaluation of relative reducts. For example, the relative reduct with the highest evaluation, 0.35, is {*milk, aquatic, backbone, fins, legs*}, and this relative reduct generates 20 decision rules; thus, we have the score $7/20 = 0.35$. In other words, because the decision table is consistent, this relative reduct divides 101 samples into 20 equivalence classes. In addition, even though the number of attributes is the same, the relative reduct with the lowest score of evaluation, 0.259, is {*eggs, aquatic, toothed, legs, catsize*}, which constructs 27 equivalence classes. Moreover, the worst relative reduct, with the evaluation score 0.167, is {*egg, aquatic, predator, breathes, venomous, legs, catsize*}, which constructs 42 equivalence classes.

Decision rules constructed from the best relative reduct {*milk, aquatic, backbone, fins, legs*} represent characteristics of each decision class very well. For example, unlike the decision class "mammal", which is directly identified by the attribute *milk*, there is no attribute that identifies the decision class "fish" directly. However, the decision class "fish", which consists of 13 kinds of fishes, is described just one decision rule:

- If (*milk* = no) and (*aquatic* = yes) and (*backbone* = yes) and (*fins* = yes) and (*legs* = 0), then (*type* = fish); Certainty = 1, Coverage = 1.

Note that the attribute *fins* alone does not identify the decision class fish, because aquatic mammals in the Zoo data such as the dolphin, porpoise, and seal also have fins. Similarly, the decision class "bird", which consists of 20 kinds of birds, is described by two decision rules:

- If (*milk* = no) and (*aquatic* = no) and (*backbone* = yes) and (*fins* = no) and (*legs* = 2), then (*type* = bird); Certainty = 1, Coverage = 0.7,
- If (*milk* = no) and (*aquatic* = yes) and (*backbone* = yes) and (*fins* = no) and (*legs* = 2), then (*type* = bird); Certainty = 1, Coverage = 0.3.

These two decision rules describe the fact that all birds in the Zoo data are characterized as non-mammal vertebrates with two legs and no fins.

8.5 Conclusions

In this chapter, we proposed a method of evaluating *relative reducts* of a given decision table by averages of the coverage of decision rules constructed from them. The proposed method is based on the roughness and correctness of partitions based on the relative reducts and evaluates the relative reducts that provide the roughest and most correct approximations to be better. Moreover, when we evaluate a relative reduct by the proposed method, we do not need to calculate actual scores of coverage of decision rules constructed from the relative reduct; we just need to know how many decision rules are generated from the relative reduct. Experimental results of the proposed method indicate that even when the number of attributes making up relative reducts is identical, the evaluation results based on the proposed method may be quite different, and decision rules generated from a good relative reduct in the sense of the proposed evaluation represent the characteristics of decision classes very well. Thus, we believe that the proposed method is very important and useful as an evaluation criterion for relative reducts.

Several issues remain to be investigated. First, we need to compare the proposed method with other evaluation criteria for relative reducts related to evaluation of partitions [11, 12, 14], in particular. The method proposed in [14], which evaluates relative reducts by the number of decision rules induced from a relative reduct. This is because the basic idea in [14] is similar to the proposed method, as described in Sect. 8.3. We also think that the idea of evaluation of subsets of condition attributes by the average coverage of decision rules is applicable to many extensions of rough set theory, such as variable precision rough sets [18] and dominance-based rough sets [3]. Thus, application of this idea to these variations of rough sets is also an interesting issue.

References

1. Bao, Y., Du, X., Deng, M., Ishii, N.: An efficient method for computing all reducts, transactions of the Japanese Society for. Artif. Intell. **19**, 166–173 (2004)
2. Forsyth, R.: Zoo data set. http://archive.ics.uci.edu/ml/datasets/Zoo (1990). Accessed 15 May 1990
3. Greco, S., Matarazzo, B., Slowinski, R.: Rough set theory for multicriteria decision analysis. Eur. J. Oper. Res. **129**, 1–47 (2001)
4. Guan, J.W., Bell, D.A.: Rough Computational Methods for Information Systems. Artif. Intell. **105**, 77–103 (1998)
5. Hedar, A.H., Wang, J., Fukushima, M.: Tabu search for attribute reduction in rough set theory. Soft Comput. **12**, 909–918 (2008)
6. Hu, F., Wang, G., Feng, L.: Fast knowledge reduction algorithms based on quick sort. In: Wang, G. et al. (eds.), Rough Sets and Knowledge Technology, LNAI 5009, pp. 72–79, Springer, New York (2008)
7. Hu, K., Diao, L. Lu, Y., Shi, C.: A Heuristic optimal reduct Algorithm. In: Leung, K.S. et al. (eds.) Intelligent Data Engineering and Automated Learning, LNCS 1983, pp. 89–99, Springer, New York (2003)

8. Pawlak, Z.: Rough sets. Int. J. Comput. Inf. Sci. **11**, 341–356 (1982)
9. Pawlak, Z.: Rough Sets: Theoretical Aspects of Reasoning about Data. Theory and Decision, Kluwer, Dordrecht (1992)
10. Polkowski, L.: Rough sets: Mathematical Foundations, Advances in Soft Computing, Physica-Verlag, Heidelberg (2002)
11. Ślęzak, D.: Normalized decision functions and measures for inconsistent decision table analysis. Fundamenta Informaticae **44**(3), 291–319 (2000)
12. Ślęzak, D.: Approximate entropy reducts. Fundamenta Informaticae **53**(3–4), 365–390 (2002)
13. Skowron, A., Rauszer, C.M.: The discernibility matrix and functions in information systems. In: Słowiński, R. (ed.) Intelligent Decision Support: Handbook of Application and Advances of the Rough Set Theory, pp. 331–362, Kluwer, Dordrecht (1992)
14. Wróblewski, J.: Adaptive Methods of Object Classification (in Polish), Ph.D. thesis, Institute of Mathematics, Warsaw University, Poland (2001)
15. Xu, J., Sun, L.: New Reduction algorithm based on decision power of decision table. In: Wang, G. et al. (eds.)Rough Sets and Knowledge Technology, LNAI 5009, pp. 180–188, Springer, New York (2008)
16. Xu, Z., Zhang, C., Zhang, S., Song, W., and Yang, B.: Efficient attribute reduction based on discernibility matrix. In: Yao, J.T. et al. (eds.) Rough Sets and Knowledge Technology, LNAI 4481, pp. 13–21, Springer, New York (2007)
17. Zhang, J., Wang, J., Li, D., He, H., Sun, J.: A new heuristic reduct algorithm based on rough sets theory. In: Dong, G. et al. (eds.) Advances in Web-Age Information Management, LNSC 2762, pp. 247–253. Springer, New York (2003)
18. Ziarko, W.: Variable precision rough set model, Journal of Computer and System. Science **46**, 39–59 (1993)

Chapter 9
Neighbor Selection for User-Based Collaborative Filtering Using Covering-Based Rough Sets

Abstract This chapter concerns Recommender systems (RSs), providing personalized information by learning user preferences. User-based collaborative filtering (UBCF) is a significant technique widely utilized in RSs. The traditional UBCF approach selects k-nearest neighbors from candidate neighbors comprised by all users; however, this approach cannot achieve good accuracy and coverage values simultaneously. We present a new approach using covering-based rough set theory to improve traditional UBCF in RSs. In this approach, we insert a user reduction procedure into the traditional UBCF approach. Covering reduction in covering-based rough sets is used to remove redundant users from all users. Then, k-nearest neighbors are selected from candidate neighbors comprised by the reduct-users. Our experimental results suggest that, for the sparse datasets that often occur in real RSs, the proposed approach outperforms the traditional UBCF, and can provide satisfactory accuracy and coverage simultaneously.

Keywords Covering-based rough sets · User-based collaborative filtering · Covering reduction · Active user · Recommender systems

9.1 Introduction

With rapid development in technology and improved economic conditions, consumers tend to demand more personalized services. Recently, *recommender systems* (RSs), which can provide item recommendations based on personal preferences that help users make purchase decisions, have become very popular [1, 2, 8].

Collaborative filtering (CF) is a significant component of the recommendation process [4, 12]. User-based collaborative filtering (UBCF) approach relies on active user neighborhood information to make predictions and recommendations [5]. *Neighborhood selection* is one crucial procedure of UBCF approach, which selects a set of users from candidate neighbors to comprise neighborhood for an active user. Whether appropriate neighborhood can be selected will have a direct bearing on the rating prediction and item recommendation. In general UBCF approach, k-nearest neighbors (k-NN) approach is proved to be the best method to generate a neighborhood,

© Springer Nature Switzerland AG 2020
S. Akama et al., *Topics in Rough Set Theory*, Intelligent Systems
Reference Library 168, https://doi.org/10.1007/978-3-030-29566-0_9

which picks the k most similar (nearest) users from candidate neighbors to comprise the neighborhood for an active user [5]. So we consider the k-NN UBCF approach as the traditional UBCF approach in the rest of this chapter.

Currently, commercial RSs have a large number of users, neighborhood must be composed of a subset of users rather than all users if RSs want to guarantee acceptable response time [5]. Accuracy measures how closed RSs predictions reflect actual user preferences, and coverage interprets the extent to which recommendations cover the set of available items. Both metrics are important in RSs. In the neighborhood of traditional UBCF approach, neighbors tend to have similar tastes, so high predicted scores from them concentrate in few types of items, even just popular items. Due to the popular items often have high ratings from users, so recommendations from the traditional UBCF approach often have high accuracy. However, types of recommendations are very limited, it leads to an unsatisfactory coverage value [3]. Therefore using the traditional UBCF is difficult to achieve good values for both metrics simultaneously.

Aiming at improving the traditional UBCF approach to obtain good values of accuracy and coverage at the same time, in this study, covering-based rough set theory is applied to RSs. We propose *covering-based collaborative filtering* (CBCF), a new approach that uses covering reduction to remove redundant users, then neighborhood is selected from candidate neighbors comprised by the reduct-users. Experimental results reveal that our proposed CBCF approach provides better recommendation results than the traditional UBCF approach.

The remainder of this chapter is organized as follows. In Sect. 9.2, some background information is provided. We review basic concepts involved in the traditional UBCF approach and covering-based rough sets, describe covering reduction algorithm. In Sect. 9.3, we analyze neighborhood selection problems, then give the detailed motivation and construction of the CBCF approach. In Sect. 9.4, we provide an example to demonstrate the CBCF approach. In Sect. 9.5, we describe our experiments and compare CBCF results with the results obtained using the traditional UBCF approach. Conclusions and suggestions for future work are presented in Sect. 9.6. Note that this paper is a revised and extended version of a previous paper [19].

9.2 Background

Here, we introduce the basic knowledge of UBCF approach, then briefly describe covering, covering approximation, and covering reduction. In addition, we analyze and compare three types of reduction algorithms.

9.2.1 Overview of the Traditional UBCF Approach

Given an RS, let U and I be finite sets of users and items. Suppose each item has the same attributes. The item attribute matrix is denoted as AM. Let $R \cup \{\star\}$ be the set of possible item rating scores, and RM be the User-Item rating matrix. Note that the absence of a rating is represented by an asterisk $(\star)$. The rating score of user u for item i is denoted $r_{u,i} \in R \cup \{\star\}$, and the average of the valid ratings of user u is denoted $\bar{r}_u$. θ is set as the threshold for rating scores, and items with $r_{u,i} \geq \theta$ are defined as items that are relevant to user u. $I_u = \{i \in I | r_{u,i} \neq \star\}$ is the set of all items rated by user u, I_u^c is the complementary set of I_u, indicates items which have not yet rated by user u.

The traditional UBCF approach is one type of CF, it utilizes the information of an active user's neighborhood to make predictions and recommendations, it can be separated into three steps:

Step 1: Similarity computation. An active user au's candidate neighbors CN_{au} are comprised all users. Based on historical rating information, compute similarity between each user $u \in CN_{au}$ and an active user au. Here, Pearson correlation coefficient approach (9.1) is popularly used as a similarity measure:

$$sim(au, u) = \frac{\sum_{i \in I_{au} \cap I_u} \left(r_{au,i} - \bar{r}_{au} \right) \left(r_{u,i} - \bar{r}_u \right)}{\sqrt{\sum_{i \in I_{au} \cap I_u} \left(r_{au,i} - \bar{r}_{au} \right)^2} \sqrt{\sum_{i \in I_{au} \cap I_u} \left(r_{u,i} - \bar{r}_u \right)^2}}, \tag{9.1}$$

where $sim(au, u)$ indicates the similarity between the active users au and user $u \in CN_{au}$. $I_{au} = \{i \in I | r_{au,i} \neq \star\}$ is the set of all items rated by active user au, $\bar{r}_{au}$ is the average rating of active user au:

$$\bar{r}_{au} = \frac{\sum_{i \in I_{au}} r_{au,i}}{card(I_{au})}. \tag{9.2}$$

Step 2: Neighborhood selection. Select the k most similar (nearest) users from CN_{au} to comprise the neighborhood $N_{au}(k)$ for the active user au;

Step 3: Rating prediction and item recommendation. Normalize ratings and according to the rating information of neighborhood, predict a rating score $p_{au,i}$ for each item i in unrated item set I_{au}^c of the active user au. The adjusted weighted sum approach (9.3) is often utilized to make rating prediction.

$$p_{au,i} = \bar{r}_{au} + \lambda \sum_{u \in N_{au}(k) \cap U_i} sim(au, u) * (r_{u,i} - \bar{r}_u), \tag{9.3}$$

where $p_{au,i}$ is the prediction of item i for active user au, $U_i = \{u \in U | r_{u,i} \neq \star\}$ is the set of users who have rated item i, and multiplier λ is a normalizing factor and is selected as

$$\lambda = \frac{1}{\sum_{u \in N_{au}(k) \cap U_i} sim(au, u)}. \tag{9.4}$$

Note that, within specific systems, these steps may overlap or the order may be slightly different. Algorithm 4 summarizes the traditional UBCF approach.

Algorithm 4 Traditional UBCF approach

Input: User-Item rating matrix RM and an active user au.
Output: Recommended items set of size N for the active user au.
   k : Number of users in the neighborhood $N_{au}(k)$ of the active user au.
   N : Number of items recommended to the active user au.
   I_{au}^c : Items which have not yet rated by the active user au.
   CN_{au} : Candidate neighbors of the active user au.
   $p_{au,i}$: Rating prediction of item i for the active user au.
1: $CN_{au} = U$, then compute similarity between active user au and each user $u \in CN_{au}$;
2: **for** each item $i \in I_{au}^c$ **do**
3:     Find the k most similar users in CN_{au} to comprise neighborhood $N_{au}(k)$;
4:     Predict rating score $p_{au,i}$ for item i by neighborhood $N_{au}(k)$;
5: **end for**
6: Recommend to the active user au the top N items having the highest $p_{au,i}$.

9.2.2 Covering-Based Rough Sets and Covering Reduction Theory

Rough set theory was first presented by Pawlak in the early 1980s [9]. Lower and upper approximation operations are key concepts in classical rough set theory, and an equivalence relation, i.e., a partition, is the simplest formulation of the lower and upper approximation operations [10]. *Covering-based rough sets* extend the classical rough set by utilizing a covering of the domain rather than a partition. Here, we define covering and covering approximation space. More detailed explanations can be found in [14, 18, 21, 22].

Definition 9.1 Let T be the domain of discourse and C be a family of subsets of T. If none of the subsets in C is empty and $\cup C = T$, C is called a covering of T.

Note that a partition of T is a covering of T; thus, the concept of a covering is an extension of a partition. Different lower and upper approximation operations generate different types of covering-based rough sets. The covering-based rough set was first presented by Zakowski [18]. Zakowski extended Pawlak's rough set theory from a partition to a covering, and presented the basic concept of covering. In addition, related studies have been undertaken by [11, 13, 15, 20, 25].

Covering reduction is a significant concept in covering-based rough set theory [16]. The concept of covering reduction was originally presented by Zhu et al. [23]. In this chapter, we refer to the algorithm proposed by Zhu et al. [20] as the first type of reduction algorithm, which corresponds to the definition of $reduct(C)$ in [20]. Definition 9.2 defines this algorithm.

Definition 9.2 Let C be a covering of domain T, and $K \in C$. If K is a union of some sets in $C - \{K\}$, K is reducible in C; otherwise, K is irreducible. When all reducible elements are removed from C, the new irreducible covering is called the first-type reduct of C.

Zhu et al. presented two other covering reduction algorithms [24, 25], which we refer to as the second and third types of reduction algorithms, respectively. Definition 9.3 defines the second-type algorithm, which corresponds to the definition of $exclusion(C)$ [24]. Definition 9.4 defines the third-type algorithm, which corresponds to the definition of $exact - reduct(C)$ [25].

Definition 9.3 Let C be a covering of domain T, and $K \in C$. If there exists another element K' of C such that $K \subset K'$, K is an immured element of covering C. When we remove all immured elements from C, the set of all remaining elements is still a covering of T, and this new covering has no immured element. We refer to this new covering as the second-type reduct of C.

Definition 9.4 Let C be a covering of domain T, and $K \in C$. If there exists $K_1, K_2...K_m \in C - K$ such that $K = K_1 \cup, ..., \cup K_m$, and $\forall x \in K$ and $\{x\}$ is not a singleton element of C, $K \subseteq \cup \{K' \mid x \in K' \in C - \{K\}\}$, K is called an exact-reducible element of C. When all exact-reducible elements are removed from C, the new irreducible covering is called the third-type reduct of C.

Comparing the three types of covering reduction algorithms, we find that, the first type removes redundant elements more efficiently than the third type because the third type has an additional restriction condition. For example, we assume that $K \in C$ is a reducible element in the first type, but if there exists $x \in K$ that $\{x\}$ is a singleton element of C, K is not an exact-reducible element in the third type. However, if $K \in C$ is an exact-reducible element in the third type, it must be a reducible element in the first type.

Here, we consider the first and second types. If we assume that $K \in C$ is a reducible element in the first-type algorithm, then there must be other elements whose union is K. For example, for $K = K_1 \cup K_2$, only K should be removed; however, under the same conditions, in the second-type algorithm, K_1 and K_2 would both be considered as immured elements, which should be removed.

Typically, an RS has a vast number of items and each user has different preferences. Therefore it is difficult to represent one user's preferred item set as a union of other users' preferred item sets accurately. In this situation, for the first-type algorithm, few reducible elements can be removed; however, for the second type, there can be a large number of reducible elements, because RSs have a large number of users, it is easy to find one user's preferred item set that includes another user's set.

Thus, the second type of covering reduction algorithm can be used to remove more reducible elements in RSs. The second-type of covering reduction algorithm (STCRA) is given in Algorithm 5.

Algorithm 5 STCRA: The second-type of covering reduction algorithm

---

Input: A covering of a domain: C.
Output: An irreducible covering of a domain: $reduct(C)$.
    K_i, K_j: Elements in the covering C.
1: set $reduct(C)=C$;
2: **for** $i = 1$ to card(C) **do**
3:    **for** $j = 1$ to card(C) **do**
4:      **if** $K_j \subset K_i$ **then**
5:        **if** $K_j \in reduct(C)$ **then**
6:          $reduct(C) = reduct(C) - \{K_j\}$;
7:        **end if**
8:      **end if**
9:    **end for**
10: **end for**
11: **return** $reduct(C)$;

9.3   Covering-Based Collaborative Filtering Approach

Here, first we discuss neighborhood selection problems in the traditional UBCF approach. To address these problems, we propose CBCF approach and describe its detail process. In addition, we discuss the innovative aspects and significance of the proposed approach.

9.3.1   Neighborhood Selection Problems in Traditional UBCF Approach

Neighborhood selection is to determine which users' rating information will be utilized to compute the prediction for an active user, in other words, it decides who will be selected as neighborhood of the active user. In theory, every user could be selected as a neighbor. However, modern commercial RSs have vast customers, e.g., Amazon has billions of users, it is impractical to consider every user as a neighbor when trying to maintain real-time performance. A subset of users must be selected as neighborhood if RSs want to guarantee acceptable response time. Herlocker et al. discussed the size of neighborhood in detail, and drew a conclusion that the size of neighborhood affects the performance of RSs in a reasonably consistent manner [5]. It suggests that, in the real-world situations, a neighborhood of 20 to 60 neighbors is reasonable to be used to make predictions.

Currently, k-NN method is often used in the traditional UBCF approach to make neighborhood selection, neighborhood is comprised by the top k users with highest similarity in candidate neighbors. However, in RSs, some items, especially the popular items, have high rating scores from most of users, and the active user usually also prefer these items. In this case, when using the traditional UBCF approach, users who prefer the popular items are likely to have high similarity with the active user, so they will easily appear in the neighborhood. Other users, who prefer niche items, are difficult to be selected as the neighborhood, but these niche items may also be preferred by the active user. For example, the relevant items of user 1 and user 2 are popular items, the relevant items of user 3 are niche items. Similarity between active user and them are 0.9, 0.8, and 0.7, respectively, besides that, user 2s relevant item set is included in user 1s relevant item set. In traditional UBCF approach, if we select two most similar users as neighborhood, user 1 and user 2 will be selected, in this case, only popular items will be recommended to the active user. However, user 3 also have high similarity with the active user, relevant items of user 3 may also be preferred by the active user. In order to obtain neighborhood with diverse tastes, we can remove user 2 and select user 1 and user 3 as the neighborhood. Because the relevant item set of user 2 is included in user 1s relevant item set, so we can only utilize user 1 to make predictions for popular items rather than both of them. Here, we consider users like user 2, whose relevant item set is included in other user's relevant item set, as the redundant users. In traditional UBCF approach, the k-nearest neighbors have similar taste, so they tend to have similar relevant items, therefore neighborhood usually contains many redundant users. When making prediction, they tend to give high predicted scores for few types of items, even just the popular items. It causes the traditional UBCF approach cannot provide recommendations with good values of accuracy and coverage simultaneously.

9.3.2 Motivation of CBCF Approach

The proposed CBCF approach aims to improve the traditional UBCF approach by reducing redundant users, and constructs neighborhood by users who have high similarity and diverse relevant items. As we discussed above, redundant user's relevant item set is included in other user's relevant item set. According to discussions in Sect. 9.2.2, reducible element in the second type of covering reduction algorithm is also included in other elements, so we can remove redundant users by using the second type of covering reduction algorithm. Removing all reducible elements means we remove all redundant users.

In general RSs, there are vast items, it means the item domain I is too large. However, different users have rated different items, it may cause not so many users could be considered as redundant users. In order to remove redundant users as many as possible, item domain should be reduced as much as possible. In CBCF approach, we reduce the domain from item set I to active user's decision class D. Items fit the active user's relevant attributes comprise the decision class D. However, in practical application, users usually do not enter their relevant attributes into RSs. Here, in order to obtain relevant attributes of an active user, we sum each attribute value in the active user's relevant item set. Due to the more high rating scores indicate that the more the active user likes the attribute, l number attributes with the largest sums are selected as the relevant attributes. Relevant attributes of the active user in the following form:

$$[at_1 = av_1] \wedge [at_2 = av_2] \wedge ... \wedge [at_m = av_m],$$

where m means the number of all attributes, at_m is an attribute and av_m is the value of at_m.

9.3.3 Construction

In CBCF approach, we insert user reduction step into the traditional UBCF approach. Algorithm 6 presents concise steps of the CBCF approach. The detailed procedure is as follows:

Step 1: User reduction. First set I as the domain, relevant items of each user comprise a set in domain I. We construct decision class D for the active user au. The decision class D consists of all items that fit the active user's relevant attributes, defined by (9.5).

$$D = \{i \in I \mid at_1(i) = av_1, at_2(i) = av_2, ..., at_m(i) = av_m\}, \tag{9.5}$$

where $at_m(i) = av_m$ means that the value of the attribute at_m on item i is av_m.

Then to remove as many redundant users as possible, we reduce the domain from item set I to decision class D, and for each user $u \in U$, the relevant items of user u in domain D comprise the relevant set C_u, where

$$C_u = \{i \in D \mid r_{u,i} \geq \theta\}. \tag{9.6}$$

Let $C^* = D - \cup C_u$; then, $C = \{C_1, C_2...C_n, C^*\}$ is a covering for the active user in domain D.

Next, based on the second type of covering reduction algorithm in the covering-based rough sets, redundant elements are removed from covering C to obtain reduct(C), we can obtain the active user's reduct-users U^r, where

$$U^r = \{u \in U \mid C_u \in reduct(C)\}. \tag{9.7}$$

Step 2: Similarity computation. Users in U^r comprise candidate neighbors CN^r_{au} of the active user au. According to the rating information, compute the similarity $sim(au, u)$ between the active user au and each user $u \in CN^r_{au}$ by the similarity measure.

Step 3: Neighborhood selection. The active user au's neighborhood $N^r_{au}(k)$ is composed by k most similar (nearest) users in CN^r_{au}.

Step 4: Rating prediction. Based on rating information of neighborhood $N^r_{au}(k)$, we predict rating score $p_{au,i}$ for each item i in unrated item set I^c_{au} of the active user au.

Algorithm 6 CBCF approach

Input: User-item rating matrix RM, item attribute matrix AM, and an active user au.
Output: Recommended items set of size N for the active user au.
  k : Number of users in the neighborhood $N_{au}^r(k)$ of the active user au.
  N : Number of items recommended to the active user au.
  D : Decision class of the active user au.
  U^r : Users after making user reduction, reduct-users.
  I_{au}^c : Items which have not yet rated by the active user au.
  CN_{au}^r : Candidate neighbors of the active user au after making user reduction.
  $p_{au,i}$: Rating prediction of item i for the active user au.
1: **for** each user $u \in U$ **do**
2:   $C_u = \{i \in D \mid r_{u,i} \geq \theta\}$.
3: **end for**
4: Let $C^* = D - \cup C_u$; then, $C = \{C_1, C_2...C_n, C^*\}$ is a covering for an active user au in domain D.
5: $reduct(C) = STCRA(C)$
6: Reduct-user $U^r = \{u \in U \mid C_u \in reduct(C)\}$.
7: $CN_{au}^r = U^r$, compute similarity between the active user au and each user $u \in CN_{au}^r$
8: **for** each item $i \in I_{au}^c$ **do**
9:   Find the k most similar users in CN_{au}^r to comprise neighborhood $N_{au}^r(k)$;
10:   Predict rating score $p_{au,i}$ for item i by neighborhood $N_{au}^r(k)$;
11: **end for**
12: Recommend to the active user au the top N items having the highest $p_{au,i}$.

9.3.4 Discussion

To provide recommendations with good values of accuracy and coverage for an active user au, the biggest innovation of the proposed CBCF approach is that, we insert the user reduction procedure into the traditional UBCF approach. For an active user au, before computing the similarity, we remove redundant users from all users to obtain reduct-users U^r and which comprise candidate neighbors CN_{au}^r with diverse tastes, k most similar (nearest) users selected from CN_{au}^r comprise neighborhood $N_{au}^r(k)$. Although comparing with input conditions of the traditional UBCF, our proposed CBCF needs an additional condition: item attribute matrix AM; however, in general RSs, item attribute matrix is very common and easy to obtain.

User reduction is a core component of CBCF approach, which applies the notion of covering reduction to reduct redundant users from all users. First, we set all items I as the domain, and relevant items of each user comprise a set in domain I. However, in this case, there are only a few sets can be removed as redundant elements. To remove as many redundant users as possible, when obtaining the decision class D, we reduce the domain from I to D such that the domain can be sufficiently small. Then, the relevant items of each user in decision class D will be a element of a covering C. Based on the definition of the second type of covering reduction algorithm, for set C_1, if there exists another set C_2 for which $C_1 \subset C_2$, C_1 is considered reducible and therefore removable. In this approach, C_1 denotes the relevant items of user 1 in domain D and $C_1 \subset C_2$ indicates that user 1 and user 2 are likely to prefer same type of items, so we can just utilize user 2 to make prediction for this type of items, thus user 1 can be considered as redundant user to be removed. Removing all reducible elements means that all redundant users are removed from all users, so that this approach can only use the reduct-users U^r to comprise CN_{au}^r. Users in CN_{au}^r have diverse relevant of items, and high similarity users are selected from CN_{au}^r to comprise neighborhood $N_{au}^r(k)$. So in proposed CBCF approach, neighbors in the $N_{au}^r(k)$ have both high similarity and diverse preference, they can make accurate predictions for more types of items and present recommendations with high accuracy and coverage at the same time.

9.4   Example of CBCF Approach in RSs

Here, we present an example to explain the CBCF approach more clearly. Table 9.1 illustrates an User-Item rating matrix RM about rating scores by six users for eight items, U_{au} represents the active user. The rating value is from 1 to 5, where a higher value indicates that the user likes the given item more.

Table 9.2 shows the item attribute matrix AM about eight items, and each item has the following attributes: Horror, Comedy, Drama, Action, and Musical, where a value of 1 indicates that the item is of that genre and a value of 0 indicates it is not. Note that items can be in several attributes simultaneously.

The detailed steps are as follow:

$Step$ 1: User reduction. Here, we treat the rating threshold θ as 3; thus, from the rating matrix RM we can obtain the active user's relevant items set {Item 2, Item 4, Item 6}. We sum each attribute value in the relevant item set according the item attribute matrix AM (Horror $= 1$, Comedy $= 3$, Drama $= 1$, Action $= 3$, Musical $= 1$). Then, two attributes with the largest sums (Comedy and Action) are selected as relevant attributes of the active user. Then, all items that fit the relevant attributes comprise the decision class $D = \{$Item 2, Item 3, Item 4, Item 5, Item 6$\}$.

Reduce the domain from all items set to decision class D. Relevant items of the user u in domain D will be a set C_u:

Table 9.1 Example of user-item rating matrix RM

User	Co-rated items						Target items	
	Item 1	Item 2	Item 3	Item 4	Item 5	Item 6	Item 7	Item 8
U_1	2	4	1	3	3	4	5	5
U_2	1	3	2	3	2	5	2	3
U_3	1	2	3	5	3	4	2	1
U_4	2	2	5	1	4	5	1	4
U_5	2	4	5	2	1	3	5	3
U_{au}	1	4	2	5	2	3	★	★

Table 9.2 Example of item attribute matrix AM

Item	Attribute				
	Horror	Comedy	Drama	Action	Musical
Item 1	1	0	1	0	0
Item 2	0	1	0	1	1
Item 3	1	1	0	1	1
Item 4	1	1	1	1	0
Item 5	0	1	1	1	0
Item 6	0	1	0	1	0
Item 7	0	1	1	0	0
Item 8	1	0	0	1	1

$C_1 = \{$Item 2, Item 4, Item 5, Item 6$\}$, $C_2 = \{$Item 2, Item 4, Item 6$\}$,
$C_3 = \{$Item 3, Item 4, Item 5, Item 6$\}$, $C_4 = \{$Item 3, Item 5, Item 6$\}$,
$C_5 = \{$Item 2, Item 3, Item 6$\}$.
Then, $C = \{C_1, C_2, C_3, C_4, C_5\}$ is a covering for the active user in domain D. Based on the definition of the second-type covering reduction algorithm, $C_2 \subset C_1$, $C_4 \subset C_3$; thus, C_2 and C_4 can be regarded as redundant elements to be removed. Then, we can obtain the reduct$(C) = \{C_1, C_3, C_5\}$, so the reduct-users $U^r = \{U_1, U_3, U_5\}$.

$Step$ 2: Similarity computation. Candidate neighbors CN_{au}^r for the active user are composed by users in U^r, $CN_{au}^r = U^r = \{U_1, U_3, U_5\}$. Then utilize the Pearson correlation coefficient similarity measure to compute the similarity between the active user and each user in CN_{au}^r. Table 9.3 shows results of similarity and user rank for the traditional UBCF and proposed CBCF approaches.

$Step$ 3: Neighborhood selection. If we consider only three nearest users in candidate neighbors as neighborhood of the active user, U_1, U_2, and U_3 will comprise the neighborhood $N_{au}(3)$ for the traditional UBCF; however, for our proposed CBCF approach, U_1, U_3, and U_5 will be considered as the neighborhood $N_{au}^r(3)$.

$Step$ 4: Rating prediction. From the rating scores of $N_{au}^r(3)$, we use the adjusted weighted sum approach to predict the rating scores for item 7 and item 8. Here $P_{au,7} = 3.284$, $P_{au,8} = 2.588$.

Because $P_{au,7} > P_{au,8}$, if we select the top one movie as recommendation, item 7 will be recommended to the active user.

9.5 Experiments and Evaluation

In this section, we introduce the evaluation dataset and metrics, examine the effects of the approach components, and compare the CBCF approach's performance with the traditional UBCF approach with different datasets.

Table 9.3 Example of similarity and rank depending on different approaches

User-User	Traditional UBCF		Proposed CBCF	
	Similarity	Rank	Similarity	Rank
$U_{au} - U_1$	0.501	3	0.501	2
$U_{au} - U_2$	0.563	2	–	–
$U_{au} - U_3$	0.646	1	0.646	1
$U_{au} - U_4$	−0.458	5	–	–
$U_{au} - U_5$	0.075	4	0.075	3

9.5.1 Experimental Setup and Evaluation Metrics

In our experiments, we utilized the MovieLens (Herlocker et al. [6]) and Jester (Ken et al. [7]) datasets because they are often used to evaluate RSs. The MovieLens 100 K dataset consists of 1,682 movies, 943 users, and 100,000 ratings on a scale of 1 to 5. Each user has rated at least 20 movies, and in our study, movies rated above 3 were treated as a user's relevant movies. The Jester 3.9M dataset contains ratings of 100 jokes from 24,983 users. Each user has rated 36 or more jokes. The value range of rating scores is −10 to 10. A value of "99" represents an absent rating. In our experiment, jokes rated above 5 were treated as a user's relevant jokes.

We also used the conventional leave-one-out procedure to evaluate the performance of the proposed approach. For each test user, we only considered items that the user had rated as test items. First, we supposed that the test items had no rating scores from the test user. Then, our approach predicted a rating score for each test item using the information obtained from the remaining users. Finally, comparisons were made between the original and predicted rating scores.

For the MovieLens dataset, we summed each attribute value in the test user's set of relevant movies, and l number attributes with the largest sums were selected as the relevant attributes of the test user. As there were 18 attributes for each movie, we computed the average size of decision class in terms of different number of l, Table 9.4 shows the result. If the size of test user's decision class is too big, there will be just fewer redundant users could be removed; however, if the size of test user's decision class is too small, other users' relevant item set will include this decision class easily, in this case, it will lose the meaning of reduction. Overall consideration, we select two attributes to construct the decision class.

For the Jester dataset, no information was presented about item attributes. There were 100 jokes in this dataset, we considered the top 50 jokes sorted by the test user's rating scores as the decision class. If the number of rated jokes from the test user was less than 50, we treated all rated jokes as the decision class. However if the neighbor's set of relevant jokes was too large, it would include the decision class, in this case, covering reduction will lose effectiveness. To avoid this, we selected the top 10% users who had rated the fewest jokes from all users, and utilized these 2,498 users for our experiment.

Table 9.4 Average size of decision class versus l with the MovieLens dataset

l(number of relevant attributes)	1	2	3	4
Average size (decision class)	583.257	70.363	7.068	0.303

To measure the performance of the proposed approach, we used the mean absolute error (MAE), root mean square error (RMSE), and coverage as evaluation metrics, all of which are popular metrics for evaluating RSs.

The MAE and RMSE metrics demonstrate the average error between predictions and real values; therefore, the lower these values, the better the accuracy of RSs.

$$MAE = \frac{1}{card(U)} \sum_{u \in U} \left(\frac{1}{card(O_u)} \sum_{i \in O_u} |p_{u,i} - r_{u,i}| \right), \qquad (9.8)$$

$$RMSE = \frac{1}{card(U)} \sum_{u \in U} \sqrt{\frac{1}{card(O_u)} \sum_{i \in O_u} (p_{u,i} - r_{u,i})^2}, \qquad (9.9)$$

where $O_u = \{i \in I \mid p_{u,i} \neq \star \wedge r_{u,i} \neq \star\}$ indicates set of items rated by user u having prediction values.

In different research fields, the coverage metric can be interpreted and defined differently. We define coverage metric as calculating the percentage of situation in which at least one k-nearest neighbors of the active user can rate an item which has not been rated by that active user. Here, let $S_{u,i}$ as the set of user u's neighbors which have rated the item i, and define $Z_u = \{i \in I \mid S_{u,i} \neq \emptyset\}$.

$$Coverage = \frac{1}{card(U)} \sum_{u \in U} \left(100 \times \frac{card(I_u^c \cap Z_u)}{card(I_u^c)} \right). \qquad (9.10)$$

In addition, the reduction rate is defined as an evaluation metric, that measures the effectiveness of removing redundant users from all users. Reduction rate is given as follows:

$$ReductionRate = \frac{1}{card(U)} \sum_{u \in U} \frac{card(CN_u - CN_u^r)}{card(CN_u)}, \qquad (9.11)$$

where CN_u means candidate neighbors of user u, CN_u^r represents user u's candidate neighbors after user reduction.

Table 9.5 Number of candidate neighbors for traditional UBCF and CBCF approaches

	UBCF	CBCF	Reduction rate
MovieLens	943	193	0.795
Jester	2,498	580	0.768

9.5.2 Experimental Results and Comparisons

We conducted experiments to demonstrate the performance of the proposed CBCF approach. In addition, using different datasets, comparisons of the CBCF and traditional UBCF approaches were performed to verify if the proposed CBCF approach could provide better recommendations or not than traditional UBCF approach. In both experiments, the Pearson correlation coefficient approach was used as the similarity measure, k-NN approach was utilized to select the neighborhood, and the adjusted weighted sum approach was used as the aggregation function. To obtain MAE, RMSR, and coverage values, according to Herlocker and Konstan [5], we selected different size k neighborhood from candidate neighbors, $k \in \{20, 25, 30, \ldots, 60\}$.

Currently, researches have gotten the conclusion that there is a trade-off relationship between accuracy and coverage in traditional UBCF approach. As increasing the size of neighborhood, coverage metric increases constantly; however, for accuracy metric, it first increases and then decreases [5, 6]. In our experiments, due to the size of neighborhood is in a small range, experimental results may appear that both accuracy and coverage increase as the size of neighborhood increases. However, it does not negate the trade-off relationship between accuracy and coverage in traditional UBCF approach.

Table 9.5 shows results about number of candidate neighbors for traditional UBCF and CBCF approaches in MovieLens and Jester datasets respectively. As can be seen, in MovieLens dataset, there are 943 users, so in traditional UBCF approach, all 943 users will be considered as candidate neighbors. After user reduction, on average, approximately 79.5% of users are removed as redundant users, so in CBCF approach, remaining 193 users will comprise the candidate neighbors.

In Jester dataset, recall that there are 2,498 users, so the number of candidate neighbors for traditional UBCF approach is 2,498. The reduction rate is 76.8%, which means approximately 76.8% of users are removed as redundant users on average, so in CBCF approach, the average number of candidate neighbors is 580.

First, we introduce comparisons between the CBCF and traditional UBCF approaches with the MovieLens dataset. Figure 9.1 shows accuracy results (MAE and RMSE) versus the size of neighborhood. As can be seen, for traditional UBCF approach, both MAE and RMSE values decrease as the size of neighborhood increases, when the size of neighborhood is 60, they obtain the least values 0.626 and 0.801 respectively.

On the other hand, for CBCF approach, the MAE and RMSE values are stable, and values of two metrics are 0.623 and 0.788 when the size of neighborhood is

Fig. 9.1 Accuracy results (MAE and RMSE) versus the size of neighborhood with MovieLens dataset

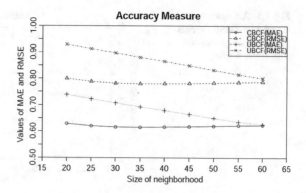

Fig. 9.2 Coverage results versus the size of neighborhood with MovieLens dataset

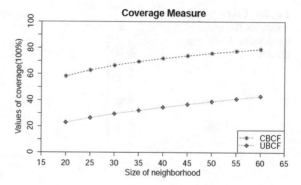

60. Overall, for MAE and RMSE metrics, all values of CBCF approach are lower than traditional UBCF approach, which means that the predicted scores by CBCF approach are closer to the original scores. So the proposed CBCF approach outperforms traditional UBCF in terms of MAE and RMSE.

Figure 9.2 illustrates the coverage metric versus the size of neighborhood. As shown in figure, the coverage of both CBCF and traditional UBCF approaches increases obviously as the size of neighborhood increases. However, the coverage of proposed CBCF approach is higher than traditional UBCF in terms of different size of neighborhood, it means, CBCF approach can recommend more types of movies that the active user has not yet rated. Thus, the comparative results for CBCF and traditional UBCF obtained with MovieLens dataset indicate that, our proposed CBCF approach can select more appropriate neighborhood, and outperform the traditional UBCF approach in terms of accuracy and coverage.

Next, we illustrate comparisons between the CBCF and traditional UBCF approaches with the Jester dataset. Figure 9.3 explains accuracy results (MAE and RMSE) versus the size of neighborhood. As shown in the figure, for both CBCF and traditional UBCF approach, values of MAE and RMSE increase slightly as the size of neighborhood increases, it means the accuracy becomes lower when the neighborhood increases. And for MAE and RMSE metrics, all values of the proposed CBCF

Fig. 9.3 Accuracy results
(MAE and RMSE) versus
the size of neighborhood
with Jester dataset

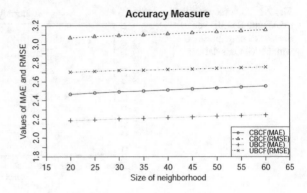

Fig. 9.4 Coverage results
versus the size of
neighborhood with Jester
dataset

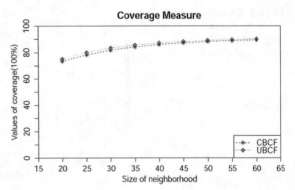

approach are higher than traditional UBCF, it indicates that CBCF approach does not outperform in terms of MAE and RMSE.

Figure 9.4 shows the coverage metric versus the size of neighborhood. As can be seen, for both CBCF and traditional UBCF approaches, coverage increases slightly as the size of neighborhood increases; however, traditional UBCF is lightly higher than the CBCF approach, which means the CBCF approach cannot recommend more types of jokes for the active user.

In conclusion, the comparative results between CBCF and UBCF with Jester dataset reveal that, the proposed CBCF approach is inferior to the traditional UBCF approach in terms of accuracy and coverage.

9.5.3 Discussion

The experimental results indicate that the proposed CBCF approach demonstrates different performance with different datasets. In the MovieLens dataset experiment, there were 1,682 movies and 943 users. For each user, the number of rated items was quite smaller than the number of unrated items; therefore, this dataset is very sparse. In the proposed CBCF approach, user reduction procedure can remove redundant users which may have high similarity but can only make predictions for few types of items, reduct-users with diverse tastes comprise the candidate neighborhood, neighborhood selected from candidate neighbors can predict rating scores for more types of items, so the coverage metric has improved greatly comparing with the traditional UBCF approach. Furthermore, in RSs, although some users have higher similarity with the active user, they cannot provide predictions with high accuracy. For example, some users have rated few items, they often also have few co-rated items with the active user, even only one; however their rating scores for co-rated items are similar. In this case, they will have high similarity, but they may not have similar preferences with the active user, so they cannot provide predictions with high accuracy. As these users have fewer rated items, in CBCF approach, they are easy to be considered as redundant users to be removed, so accuracy metric of CBCF approach has a great improvement than traditional UBCF approach.

In the Jester dataset experiment, we utilized 2,498 users; however, this dataset has only 100 jokes. Thus, for each user, there are fewer unrated jokes than rated jokes. Each joke may be rated many times by different users; thus, this dataset is not sparse. Under these circumstances, all jokes can be considered as popular jokes, and each user can predict rating scores for sufficient types of jokes relative to all 100 jokes. Due to co-rated items are sufficient between each two users, so users having higher similarity with the active user can also provide predictions with higher accuracy. In CBCF approach, user reduction procedure removes some redundant users with higher similarity; however, these users can make predictions with higher accuracy, so the accuracy metric decreases comparing with UBCF approach. Besides, as there are only 100 jokes, and each user has rated sufficient jokes, it means each user can make predictions for almost same types of jokes. So after user reduction, reduct-users, which comprise candidate neighbors, may not have improvements to make predictions for more types of jokes. Therefore, comparing with traditional UBCF approach, the coverage metric of CBCF approach does not have improvements.

Generally, in practical applications, RSs must handle big data that include huge numbers of users and items. Thus, for each user, only small number of items have been rated compared to the huge number of unrated items. Thus, most RSs have sparse datasets, such as the MovieLens dataset. However, for a sparse dataset, the proposed CBCF approach can select more appropriate neighborhood than the UBCF approach and can make recommendations for the active user with satisfactory accuracy and coverage values simultaneously. Thus, the proposed CBCF approach has important significance for RSs.

9.6 Conclusion and Future Work

UBCF approach is the most commonly used and studied technology for making recommendations in RSs. Generally, we use accuracy and coverage to evaluate an RS; however, although neighborhood selected by the traditional UBCF approach has high similarity with the active user, neighborhood tends to have similar tastes, so they are like to give high rating scores for few types of items, even only the popular items. Therefore it is difficult for the traditional UBCF approach to provide satisfactory accuracy and coverage simultaneously.

In this chapter, we have presented the CBCF approach based on covering-based rough sets to improve the traditional UBCF approach. In the proposed CBCF approach, we add the user reduction procedure into the traditional UBCF, covering reduction in covering-based rough set is utilized to remove redundant users from all users, users having diverse preferences comprise reduct-users. Neighborhood is composed by k most similar users in candidate neighbors which consist of reduct-users, so that neighbors in the neighborhood not only have high similarity but also have diverse tastes. Our experimental results indicate that, for sparse datasets (which often appears in practical RSs), unlike traditional UBCF, the proposed CBCF approach can provide recommendations with good values of accuracy and coverage simultaneously. Thus, the proposed CBCF approach can recommend satisfactory recommendations and obtain high confidence from the active user.

In the future, we plan to improve the proposed CBCF approach to address the new user cold-start problem, which is an extremely difficult issue in RSs; however, the proposed CBCF approach cannot select appropriate neighborhood based on insufficient new user data. We must propose a new similarity measure to select neighborhood efficiently and make satisfactory recommendations for new users.

References

1. Adomavicius, G., Tuzhilin, A.: Toward the next generation of recommender systems: a survey of the state-of-the-art and possible extensions. IEEE Trans. Knowl. Data Eng. **17**, 734–749 (2005)
2. Bobadilla, J., Ortega, F.: Recommender system survey. Knowl.-Based Syst. **46**, 109–132 (2013)
3. Gan, M.X., Jiang, R.: Constructing a user similarity network to remove adverse influence of popular objects for personalized recommendation. Expert Syst. Appl. **40**, 4044–4053 (2013)
4. Hameed, M.A., Jadaan, O.A., Ramachandram, S.: Collaborative filtering based recommendation system: A survey. Int. J. Comput. Sci. Eng. **4**, 859–876 (2012)
5. Herlocker, J.L., Konstan, J.A.: An empirical analysis of design choices in neighborhood-based collaborative filtering algorithms. Inf. Retr. **5**, 287–310 (2002)
6. Herlocker, J.L., Konstan, J.A., Borchers, A., Riedl, J.: An algorithmic framework for performing collaborative filtering. In: Proceedings of the 22nd Annual International ACM SIGIR Conference on Research and Development in Information Retrieval, pp. 230–237 (1999)
7. Ken, G., Theresa, R., Dhruv, G., Chris, P.: Eigentaste: a constant time collaborative filtering algorithm. Inf. Retr. **4**, 133–151 (2001)
8. Lu, J., Wu, D., Mao, M.: Recommender system application developments: a survey. Decis. Support Syst. **74**, 12–32 (2015)
9. Pawlak, Z.: Rough sets. Int. J. Comput. Inf. Sci. **11**, 341–356 (1982)
10. Pawlak, Z., Skowron, A.: Rudiments of rough sets. Inf. Sci. **177**, 3–27 (2007)
11. Pomykala, J.A.: Approximation operations in approximation space. Bull. Pol. Acad. Sci. Math. **35**, 653–662 (1987)

12. Symeonidis, P., Nanopoulos, A., Papadopoulos, A.N., Manolopoulos, Y.: Collaborative recommender systems: combining effectiveness and efficiency. Expert Syst. Appl. **34**, 2995–3013 (2008)
13. Tsang, E., Cheng, D., Lee, J., Yeung, D.: On the upper approximations of covering generalized rough sets. In: Proceedings of the 3rd International Conference Machine Learning and Cybernetics, pp. 4200–4203, (2004)
14. Tsang, E., Chen, D., Yeung, D.S.: Approximations and reducts with covering generalized rough sets. Comput. Math. Appl. **56**, 279–289 (2006)
15. Wang, J., Dai, D., Zhou, Z.: Fuzzy covering generalized rough sets. J. Zhoukou Teach. Coll. **21**, 20–22 (2004)
16. Yang, T., Li, Q.G.: Reduction about approximation spaces of covering generalized rough sets. Int. J. Approx. Reason. **51**, 335–345 (2010)
17. Yao, Y.Y., Yao, B.X.: Covering based rough set approximations. Inf. Sci. **200**, 91–107 (2012)
18. Zakowski, W.: Approximations in the space (u, π). Demonstr. Math. **16**, 761–769 (1983)
19. Zhang, Z.P., Kudo, Y., Murai, T.: Applying covering-based rough set theory to user-based collaborative filtering to enhance the quality of recommendations. In: Proceedings of the 4th International Symposium on IUKM, pp. 279–289 (2015)
20. Zhu, W.: Topological approached to covering rough sets. Inf. Sci. **177**, 1499–1508 (2007)
21. Zhu, W.: Relationship between generalized rough sets based on binary relation. Inf. Sci. **179**, 210–225 (2009)
22. Zhu, W.: Relationship among basic concepts in covering-based rough sets. Inf. Sci. **179**, 2478–2486 (2009)
23. Zhu, W., Wang, F.Y.: Reduction and maximization of covering generalized rough sets. Inf. Sci. **152**, 217–230 (2003)
24. Zhu, W., Wang, F.Y.: On three types of covering-based rough sets. IEEE Trans. Knowl. Data Eng. **19**, 1131–1144 (2007)
25. Zhu, W., Wang, F.Y.: The fourth type of covering-based rough sets. Inf. Sci. **201**, 80–92 (2012)

Chapter 10
Granular Computing and Aristotle's Categorical Syllogism

Abstract This chapter applies granular computing to Aristotle's categorical syllogism. Such kind of reasoning is called granular reasoning. For the purpose, two operations called zooming in and out are introduced to reconstruct granules of possible worlds.

Keywords Granular reasoning · Zooming in and out · Filtration · Aristotle's syllogism

10.1 Introduction

Recently, Lin [3], Skowron [10], and others have developed *granular computing* based on *rough set theory* (Pawlak [7, 8]) and many researchers expect that it provides a new paradigm of computing. In this chapter, by granular reasoning, we mean some mechanism for reasoning using granular computing. We described in [4] a possible step for granular reasoning using *filtration* in modal logic [2]. Then, in [5], we applied the idea to propositional reasoning using two operations called there *zooming in and out*. This chapter aims to provide the next step for formulating granularity of an aspect of Aristotle's *syllogism*.

But why now Aristotle? In our opinion, Aristotle's categorical syllogism can capture the essence of human ordinary reasoning, not of mathematical reasoning. We cannot forget the fact that the great work of Frege provides a precise expression of mathematical inference, but when it was applied to human ordinary reasoning, we had to face some very difficult problems including the frame problem. Frege analyzed a universal sentence '(All) s is p' using his invention universal quantifier as $\forall x(s(x) \rightarrow p(x))$.

Frege's analysis is undoubtedly correct and was the motive power of a great deal of brilliant results in mathematical logic since the 20th century. At the same time, nevertheless, his analysis was too much detailed for human beings, in general, to carry out their ordinary reasoning in his fashion. It is well-known that Frege analyzed sentences in a structure of 'individual-predicate,' while Aristotle's base structure was 'subject-predicate'.

© Springer Nature Switzerland AG 2020
S. Akama et al., *Topics in Rough Set Theory*, Intelligent Systems
Reference Library 168, https://doi.org/10.1007/978-3-030-29566-0_10

Clearly, however, we do not have to analyze every concept at the level of individuals every time for ordinary reasoning. Frege requires us complete analysis, which may cause intractability, while Aristotle's logic seems to base granules.

This chapter aims to give a step for describing Aristotle's syllogism in a way of granular computing. Let $M = \langle U, I \rangle$ be a structure for predicate logic, where U is a domain of individuals and I is an interpretation. In rough set theory, sentences in predicate logic is often represented in a Kripke-style model M, where a domain and predicate symbols are regarded as a set of worlds and atomic sentences, respectively, and thereby we have, for example, the following correspondence

$$M \models mortal(socrates) \text{ iff } M, socrates \models mortal. \qquad (10.1)$$

Here make a quotient model $M/\sim_{human}$ using an equivalence relation $\sim_{human}$ which means two worlds are equivalent just in case they both have the property '*human*' in common. Then one equivalent class must be the set of humans denoted by $I(human)$, where I is an interpretation in predicate logic.

For example, $I(human) = [socrates]_{\sim_{human}}$ holds by taking *socrates* as its representative element. Thus, in the quotient model, for instance, we may write

$$M/\sim_{human}, I(human) \models mortal \qquad (10.2)$$

Here can we see this expression (10.2) as corresponding to Aristotle's analyses '(All) human is mortal'? This may suggest a possible way of formulating higher-order predicate expressions like $mortal(human)$ under, for instance, the following correspondence: for some structure M',

$$M' \models mortal(human) \text{ iff } M/\sim_{human}, I(human) \models mortal,$$

which is parallel to formula (10.1). This is the starting point for our research. Then, in such an approach, we need reasoning process using reconstruction of models. We call such operations 'zooming in and out' in [4], which is defined based on some idea of granularity introduced to give a logical foundation of local and global worlds in semantic fields [9].

Such operators provide us a way of changing our point of view moving from global to local and vice versa. In this chapter, using such operations of 'zooming in and out', we try to describe Aristotle's categorical syllogism as a small step for formulating granular reasoning.

10.2 Granules of Possible Worlds

Given a countably infinite set of atomic sentences $\mathcal{P}$, a language $\mathcal{L}_{BL}(\mathcal{P})$ for propositional logic of belief is formed as the least set of sentences from $\mathcal{P}$ with the well-known set of connectives with a modal operator B (belief) by the usual formation rules. A sentence is called *non-modal* if it does not contain any occurrence of B.

A Kripke model is a tuple $M = \langle U, R, V \rangle$, where U is a non-empty set of (possible) worlds, R is a binary relation on U, and V is a valuation for every atomic sentence p at every world x. Define $M, x \models p$ iff $V(p, x) = 1$. The relationship $\models$ is extended for every compound sentence in the usual way. The *truth set* of p in M is defined as $\|p\|^M = \{x \in U \mid M, x \models p\}$. A sentence p is said to be *valid* in M, written $M \models p$, just in case $\|p\|^M = U$.

Given $\mathcal{P}$, we call any subset in x an *elementary world* in $U = 2^{\mathcal{P}}$. Then, for an atomic sentence p in $\mathcal{P}$ and an elementary world x in U, a valuation V is naturally defined by $V(p, x) = 1$ iff $p \in x$. When a binary relation R is given on U, we have a Kripke model $M = \langle U, R, V \rangle$. The set U is, in general, large for us when we are concerned only with finite sentences in ordinary inference. Hence we need some way of granularizing U.

Our proposal in this chapter is to make a quotient set whose elements are regarded as granules of possible worlds. Suppose we are concerned with a set Γ of non-modal sentences. Let $\mathcal{P}_\Gamma = \mathcal{P} \cap \mathrm{sub}(\Gamma)$, where $\mathrm{sub}(\Gamma)$ is the union of the sets of subsentences of each sentence in Γ.

Then, we can define an agreement relation $\sim_\Gamma$ by

$$x \sim_\Gamma y \text{ iff } \forall p \in \mathcal{P}_\Gamma (V(p, x) = V(p, y))$$

The relation becomes an equivalence relation and induces the quotient set

$$U_\Gamma =_{def} U / \sim_\Gamma .$$

We regard its (non-empty) elements as the granules of possible worlds under Γ. A new valuation is given by $V_\Gamma(p, X) = 1$ iff $p \in \cap X$, for p in $\mathcal{P}_\Gamma$ and X in U_Γ.

According to Chellas [2], when a relation R is given on U, assume we have an accessibility relation R' on U_Γ satisfying (a) if xRy then $[x]_{\sim_\Gamma} R'[y]_{\sim_\Gamma}$, (b) if $[x]_{\sim_\Gamma} R'[y]_{\sim_\Gamma}$ then $M, x \models \mathrm{B}p \Rightarrow M, y \models p$, for every sentence $\mathrm{B}p$ in Γ, and (c) if $[x]_{\sim_\Gamma} R'[y]_{\sim_\Gamma}$ then $M, y \models p \Rightarrow M, x \models \neg\mathrm{B}\neg p \models p$, for every sentence $\neg\mathrm{B}\neg p$ in Γ, then the model $M_\Gamma^{R'} = \langle U_\Gamma, R', V_\Gamma \rangle$ is called a *filtration* through $\mathrm{sub}(\Gamma)$.

10.3   Zooming In and Out

10.3.1   Zooming In and Out on Sets of Worlds

Let Γ be a set of non-modal sentences we are concerned with at a given time. It is called a *focus* at the time. When we move our viewpoint from one focus to another along time, we must reconstruct the set of granularized possible worlds. Let Γ be the current focus and Δ be the next focus we will move to. First we consider the simpler two nested cases.

(a) When $\mathcal{P}_\Gamma \supseteq \mathcal{P}_\Delta$. We need granularization, which is represented by a mapping $\mathcal{I}_\Delta^\Gamma : U_\Gamma \to U_\Delta$, called a *zooming in from* Γ to Δ, where, for any X in U_Γ,

$$\mathcal{I}_\Delta^\Gamma(X) \stackrel{\text{def}}{=} \{x \in X \mid x \cap \mathcal{P}_\Delta = (\cap X) \cap \mathcal{P}_\Delta\}.$$

(b) When $\mathcal{P}_\Gamma \subseteq \mathcal{P}_\Delta$. We need an inverse operation of granularization $O_\Delta^\Gamma : U_\Gamma \to 2^{U_\Delta}$, called a *zooming out from* Γ to Δ, where, for any X in U_Γ,

$$O_\Delta^\Gamma(X) \stackrel{\text{def}}{=} \{Y \in U_\Delta \mid (\cap Y) \cap \mathcal{P}_\Gamma = \cap X\}$$

Finally, for non-nested two sets $\mathcal{P}_\Gamma, \mathcal{P}_\Delta$, the movement from Γ to Δ can be represented using combination of 'zooming out and in' as $\mathcal{I}_\Delta^{\Gamma \cup \Delta} \circ O_{\Gamma \cup \Delta}^\Gamma : U_\Gamma \to 2^{U_\Delta}$.

10.3.2   Extending Zooming In and Out on Models

We extend the two operations so that they can be applied to models. Again let Γ and Δ be the current and the next focus, respectively. Given a model $\mathcal{M}_\Gamma = \langle U_\Gamma, R_\Gamma, V_\Gamma \rangle$, we define, for X in U_Γ, $P_\Gamma(X) \stackrel{\text{def}}{=} \{X' \in U_\Gamma \mid X R_\Gamma X'\}$. We abbreviate it to P_Γ, when $P_\Gamma(X) = P_\Gamma(X')$ for every X, X' in U.

(a) When $\Gamma \supseteq \Delta$, a *zooming in of* $\mathcal{M}_\Gamma$ through Δ is a tuple

$$\mathcal{I}_\Delta^\Gamma(\mathcal{M}_\Gamma) \stackrel{\text{def}}{=} \langle U_\Delta, R_\Delta, V_\Delta \rangle,$$

where $Y_i R_\Delta Y_j$ iff $Y_j \in \mathcal{I}_\Delta^\Gamma(\cup \{P_\Gamma(X) \mid X \in U_\Gamma \text{ and } \mathcal{I}_\Delta^\Gamma(X) = Y_i\})$.

(b) When $\Gamma \subseteq \Delta$, a *zooming out of* $\mathcal{M}_\Gamma$ through Δ is a tuple

$$O_\Delta^\Gamma(\mathcal{M}_\Gamma) \stackrel{\text{def}}{=} \langle U_\Delta, R_\Delta, V_\Delta \rangle,$$

where $Y_i R_\Delta Y_j$ iff $Y_j \in O_\Delta^\Gamma(P_\Gamma((O_\Delta^\Gamma)^{-1}(Y_i)))$.

(c) Non-nested cases are described using a *merging* of two models $\mathcal{M}_\Gamma$ and $\mathcal{M}_\Delta$.

When $\mathcal{M}_\Gamma = \mathcal{M}_\Delta$, their merging $\mathcal{M}_\Gamma \circ \mathcal{M}_\Delta$ is $\langle U_\Gamma, R, V_\Gamma \rangle$, where $X_i R X_j$ iff $X_j \in P_\Gamma(X_i) \cap P_\Delta(X_i)$.

The merging $\circ$ is extended for the cases $\mathcal{M}_\Gamma \neq \mathcal{M}_\Delta$: If $\mathcal{M}_\Gamma \supset \mathcal{M}_\Delta$, then $\mathcal{M}_\Gamma \circ \mathcal{M}_\Delta \stackrel{\text{def}}{=} \mathcal{M}_\Gamma \circ O_\Gamma^\Delta(\mathcal{M}_\Delta)$, else if $\mathcal{M}_\Gamma \subset \mathcal{M}_\Delta$, then $\mathcal{M}_\Gamma \circ \mathcal{M}_\Delta \stackrel{\text{def}}{=} O_\Gamma^\Delta(\mathcal{M}_\Gamma) \circ \mathcal{M}_\Delta$, else $\mathcal{M}_\Gamma \circ \mathcal{M}_\Delta \stackrel{\text{def}}{=} O_{\Gamma \cup \Delta}^\Gamma(\mathcal{M}_\Delta)$. The last case of merging is used for non-nested cases.

10.4 Aristotle's Syllogism and Granularity

We confine ourselves to a monadic predicate logic. Let $\mathcal{P}$ be a non-empty set of predicate symbols of arity 1 with $\top$ and $\bot$. Let C be a non-empty set of constants. For a structure $M = \langle U, I \rangle$ for a monadic predicate logic, where U is a domain and I is an interpretation.

Define a mapping $\varphi : U \to \mathbf{2}$ by $\varphi(x) = \{p \in \mathcal{P} \mid I(x) \in I(p)\}$. Since, when φ is not injective, we can replace U by U/φ, for simplicity, if we assume φ is an injection, thus any individual can be identified with its corresponding element in $\mathbf{2}$.

Then, we can formulate a model for logic of relative modality (cf. Chellas [2]) as $M = \langle U, \{B^p\}_{p \in \mathcal{B}}, V \rangle$, where $B^p = I(p)$ and V is a valuation defined by $V(p, a) = 1$ iff $a \in I(p)$ for a predicate symbol p and a in U. From B^p, we can recover a binary relation R^p on U by $a R^p b$ iff $b \in B^p$, and thus M is a Kripke model. Note that $\|p\|^M = I(p)$. For an atomic sentence $p(c)$ in a monadic predicate logic, we have obviously

$$M \models p(c) \text{ iff } M, a \models p,$$

where $I(c) = a$.

For two predicate symbols s and p, we have the following two lemmas.

Lemma 10.1 $M \models [s]p$ iff $B^s \subseteq \|p\|^M$ iff $I(s) \subseteq I(p)$.

Lemma 10.2 $M \models \langle s \rangle p$ iff $B^s \cap \|p\|^M \neq \emptyset$ iff $I(s) \cap I(p) \neq \emptyset$.

For simplicity, we assume that $I : C \to U$ is a bijection, thus we identify any constant with its corresponding element $I(c)$ in U.

Consider representation of the four basic types of sentences in a Kripke model.

Universal sentence $\begin{cases} \mathbf{A} : \text{All } s \text{ is } p. \\ \mathbf{E} : \text{No } s \text{ is } p \end{cases}$

Particular sentences $\begin{cases} \mathbf{I} : \text{Some } s \text{ is } p. \\ \mathbf{O} : \text{Some } s \text{ is not } p \end{cases}$

10.4.1 Universal Sentences and Lower Zooming In

First we consider translation of a universal sentence of type **A** like 'All human is mortal' into the above kind of Kripke models. Since Frege's achievement, it is well known that a universal sentence 'All s is p' is translated into a predicate logic as $\forall x(s(x) \to p(x))$. Given $M = \langle U, I \rangle$, because $M \models \forall x(s(x) \to p(x))$ iff $I(s) \subseteq I(p)$, we have $M \models \forall x(s(x) \to p(x))$ iff $M \models [s]p$, and thus

Lemma 10.3 'All s is p' is true iff $M \models [s]p$.

Consider a zooming in of M through $\{s, p\}$:

$$M_{s,p} \overset{\text{def}}{=} I_{s,p}(M) = \langle U_{s,p}, \{B^q_{s,p}\}_{,q\in\mathcal{B}}, V_{s,p}\rangle.$$

Note that $B^q_{a,p} = I_{s,p}(B^q)$. In general, $U_{s,p} = \{X_1, X_2, X_3, X_4\}$, where $X_1 = I(s) \cap I(p)$, $X_2 = I(s) \cap I(p)^c$, $X_3 = I(s)^c \cap I(p)$, $X_4 = I(s)^c \cap I(p)^c$. If $I(s) \subseteq I(p)$, then $X_2 = \emptyset$, and thus $U_{s,p} = \{X_1, X_3, X_4\}$. Then, $B^s_{s,p} = \{X_1\} \subseteq \{X_1, X_3\} = \|p\|^{s,p}$. The converse is also shown. Hence we have $M \models [s]p$ iff $M_{s,p} \models [s]p$ and thus

Lemma 10.4 *'All s is p' is true iff $M_{s,p} \models [s]p$.*

Consider further zooming in. We make a quotient set $U_s = U/R^s = \{I(s), I(s)^c\}$. When $I(s) \subseteq I(p)$, we have $I(s) \subseteq R^s(I(p))$. And thus we can construct the *lower zooming in* from M_s using relative filtration [4].

Definition 10.1 The *lower zooming in* of $M_{s,p}$ through $\{s\}$ is a tuple

$$\underline{M_s} \overset{\text{def}}{=} \underline{I^{s,p}_s}(M_{s,p}) = \langle U_s, \{B^q_s\}_{q\in\mathcal{B}}, \underline{V_s}\rangle,$$

where $\underline{V_s}(p, X) = 1$ iff $X \subseteq R^s(I(p))$, for $X \in U_s$.

Lemma 10.5 *'All s is p' is true iff $\underline{M_s}, I(s) \models p$.*

Example 10.1 Let us consider a structure $\langle U, I\rangle$where 'All human is mortal', that is, $\forall x(human(x) \rightarrow mortal(x))$ is true, which means that, in the structure, $I(human) \subseteq I(mortal)$ holds. We abbreviate *human* and *mortal* as h and m, respectively. We construct a Kripke model $M = \langle U, \{B^q\}_{q\in\mathcal{B}}, V\rangle$, where V is given, for instance, by the left-hand side of table in Fig. 10.1.

Since $B^h = I(human) \subseteq I(mortal)$, we have $M \models [human]mortal$. Next, we make a zooming in of M through $\{h, m\}$ as $M_{h,m} = \langle U_{h,m}, \{B^q_{h,m}\}_{q\in\mathcal{B}}, V_{h,m}\rangle$, where $U_{h,m} = \{[a_i]_{h,m}, [b_i]_{h,m}, [c_i]_{h,m}\}$ and $V_{h,m}$ is given in the left side table in Fig. 10.1. Note that $B^h_{h,m} = \{[a_i]_{h,m}\}$ and $\|mortal\|^{h,m} = \{[a_i]_{h,m}, [b_i]_{h,m}\}$.

Then, we $M_{h,m} \models [human]mortal$. Finally we make the lower zooming in of $M_{h,m}$ through $\{h\}$ as $\underline{M_h} = \langle U_h, \{B^q_h\}_{q\in\mathcal{B}}, \underline{V_h}\rangle$, where $U_h = \{I(human), I(human)^c\}$ with $I(human) = [a_i]_{h,m}$ and $I(human)^c = [b_i]_{h,m} \cup [c_i]_{h,m}$ and $\underline{V_h}$ is given by the right-hand side table in Fig. 10.2. Since $R^h(I(mortal)) = [a_i]_{h,m} = I(human)$, we have $\underline{M_h}, I(human) \models mortal$.

Lemma 10.6 *'No s is p' is true iff $M \models [s]\neg p$.*

Lemma 10.7 *'No s is p' is true iff $M_{s,p} \models [s]\neg p$ (Zooming in).*

Lemma 10.8 *'No s is p' is true iff $\underline{M_s}, I(s) \models \neg p$ (Lower zooming in).*

$\mathcal{M}$		human	mortal	$\cdots$
U	a_i	1	1	
		$\cdots$		
	a_j	1	1	
	b_i	0	1	
		$\cdots$		
	b_j	0	1	
	c_i	0	0	
		$\cdots$		
	c_j	0	0	

$$\mathcal{I}_{h,m} \longrightarrow \quad \text{Zooming in}$$

$\mathcal{M}_{h,m}$		human	mortal	$\cdots$
$U_{h,m}$	$[a_i]_{h,m}$	1	1	
	$[b_i]_{h,m}$	0	1	
	$[c_i]_{h,m}$	0	0	

Fig. 10.1 Zooming in of $\mathcal{M}$ through $\{human, mortal\}$

$\mathcal{M}_{h,m}$		human	mortal	$\cdots$
$U_{h,m}$	$[a_i]_{h,m}$	1	1	
	$[b_i]_{h,m}$	0	1	
	$[c_i]_{h,m}$	0	0	

$$\mathcal{I}_h^{h,m} \longrightarrow \quad \text{Lower zooming in}$$

$\mathcal{M}_h$		human	mortal	$\cdots$
U_h	$I(human)$	1	1	
	$I(human)^C$	0	0	

Fig. 10.2 Lower zooming in of $\mathcal{M}_{h,m}$ through $\{h(uman)\}$

10.4.2 Particular Sentences and Upper Zooming In

Next we consider translation of a particular sentence of type **I** like 'Some human is genius' into the kind of Kripke models. Also, since Frege, it is well known that 'Some s is p' is translated into a predicate logic as $\exists x(s(x) \wedge p(x))$. Because $\mathcal{M} \models \exists x(s(x) \wedge p(x))$ iff $I(s) \cap I(p) \neq \emptyset$, we have $\mathcal{M} \models \exists x(s(x) \wedge p(x))$ iff $\mathcal{M} \models \langle s \rangle p$, and thus

Lemma 10.9 *'Some s is p' is true iff $\mathcal{M} \models \langle s \rangle p$.*

For a zooming in of $\mathcal{M}$ through $\{s, p\}$ we have $\mathcal{M} \models \langle s \rangle p$ iff $\mathcal{M}_{s,p} \models \langle s \rangle p$.

Lemma 10.10 *'Some s is p' is true iff $\mathcal{M}_{s,p} \models \langle s \rangle p$.*

Again, let us consider further zooming in. Here we make a quotient set $U_s = U/R^S = \{I(s), I(s)^c\}$, then by $I(s) \cap I(p) \neq \emptyset$, we have $I(s) \subseteq \overline{R^S}(I(p))$, and then, we construct the upper zooming in of $\mathcal{M}_{s,p}$ through $\{s\}$.

Definition 10.2 The *upper zooming in* of $\mathcal{M}_{s,p}$ through $\{s\}$ is a tuple

$$\overline{\mathcal{M}_s} \overset{\text{def}}{=} \overline{\mathcal{I}_s^{s,p}}(\mathcal{M}_{s,p}) = \langle U_s, \{B_s^q\}_{q \in \mathcal{B}}, \overline{V_s} \rangle,$$

$\mathcal{M}$		human	genius	$\cdots$
U	a_i	1	1	
			$\cdots$	
	a_j	1	1	
	b_i	1	0	
			$\cdots$	
	b_j	1	0	
	c_i	0	1	
			$\cdots$	
	c_j	0	1	
	d_i	0	0	
			$\cdots$	
	d_j	0	0	

$$\Im_{h,g} \atop \xrightarrow{\hspace{1cm}} \atop \text{Zooming in}$$

$\mathcal{M}_{h,g}$		human	genius	$\cdots$
$U_{h,g}$	$[a_i]_{h,g}$	1	1	
	$\lfloor b_i \rfloor_{h,g}$	1	0	
	$[c_i]_{h,g}$	0	1	
	$[d_i]_{h,g}$	0	0	

Fig. 10.3 Zooming in of $\mathcal{M}$ through $\{human, genius\}$

where $\overline{V_s}(p, X) = 1$ iff $X \subseteq \overline{R^s}(I(p))$, for $X \in U_s$.

Lemma 10.11 *'Some s is p' is true iff $\overline{\mathcal{M}_s}, \models p$.*

Example 10.2 Consider a structure $\langle U, I \rangle$, where 'Some human is genius', i.e., $\exists x (human(x) \wedge genius(x))$ is true, which means, in the structure, $I(human) \cap I(genius) \neq \emptyset$ holds. Then we construct a Kripke model $\mathcal{M} = \langle U, \{B^q\}_{q \in \mathcal{B}}, V \rangle$, where V is given, for instance, by the left-hand side table in Fig. 10.3.

Since $B^h \cap \|genius\|^{\mathcal{M}} \neq \emptyset$, we have $\mathcal{M} \models \langle human \rangle genius$. For a zooming in of $\mathcal{M}$ through $\{h, g\}$ as $\mathcal{M}_{h,g} = \langle U_{h,g}, \{\mathcal{B}^q_{h,g}\}_{q \in \mathcal{B}}, V_{h,g} \rangle$, where $U_{h,g} = \{[a_i]_{h,g}, [b_i]_{h,g}, [c_i]_{h,g}\}$, and $V_{h.g}$ is given by the right-hand side table in Fig. 10.3.

Because $B^h_{h,g} = \{[a_i]_{h,g}, [b_i]_{h.g}\}$ and $\|genius\|^{h,g} = \{[a_i]_{h,g}, [c_i]_{h,g}, \}$, we have $\mathcal{M}_{h,g} = \langle human \rangle genius$.

Finally for the upper zooming in of $\mathcal{M}_{h,g}$ through $\{h\}$, i.e., $\overline{\mathcal{M}_h} = \langle U_h, \{B^q_h\}_{q \in \mathcal{B}}, \overline{V}_h \rangle$, where $U_h = \{I(human), I(human)^c\}$ with $I(human) = [a_i]_{h,g} \cup [c_i]_{h,g}$ and $I(human)^c = [c_i]_{h,g} \cup [d_i]_{h,g}$ and $\overline{V}_h$ is given by the right-hand side table in Fig. 10.4. Hence we have $\overline{\mathcal{M}_h}, I(human) \models genius$.

For a universal sentence of type **O** like 'Some human is not genius', we have similar results:

Lemma 10.12 *'Some s is not p' is true iff $\mathcal{M} \models \langle s \rangle \neg p$.*

Lemma 10.13 *'Some s is not p' is true iff $\mathcal{M}_{s,p} \models \langle s \rangle \neg p$. (Zooming in)*

Lemma 10.14 *'Some s is not p' is true iff $\overline{\mathcal{M}_s}, I(s) \models \neg p$. (Upper zooming in)*

$\mathcal{M}_{h,g}$		human	genius	$\cdots$
$U_{h,g}$	$[a_i]_{\flat,g}$	1	1	
	$[b_i]_{h,g}$	1	0	
	$[c_i]_{h,g}$	0	1	
	$[d_i]_{h,g}$	0	0	

$\mathfrak{I}_h^{h,g}$

$\longrightarrow$

(Upper zooming in)

$\mathcal{M}_h$		human	genius	$\cdots$
U_h	$I(human)$	1	1	
	$I(human)^C$	0	1	

Fig. 10.4 Upper zooming in of $\mathcal{M}_{h,g}$ through $\{human\}$

10.4.3 Conversion

Representation of several conversion rules is trivial:

Some s is p.   iff Some p is s.
$$\mathcal{M}_{s,p} \models \langle s \rangle p \text{ iff } \mathcal{M}_{s,p} \models \langle p \rangle s.$$
$$\mathcal{M}_s, I(s) \models p \text{ iff } \overline{\mathcal{M}}_p, I(p) \models s.$$

No s is p.   iff No p is s.
$$\mathcal{M}_{s,p} \models [s]\neg p \text{ iff } \mathcal{M}_{s,p} \models [p]\neg s.$$
$$\underline{\mathcal{M}}_s, I(s) \models \neg p \text{ iff } \underline{\mathcal{M}}_p, I(p) \models \neg s.$$

Not (All s is p). iff Some s is not p.
$$\mathcal{M}_{s,p} \models \neg[s]p \text{ iff } \mathcal{M}_{s,p} \models \langle s \rangle \neg p.$$
$$\underline{\mathcal{M}}_s, I(s) \not\models p \text{ iff } \overline{\mathcal{M}}_s, I(s) \models \neg p.$$

Not (Some s is p). iff No s is p.
$$\mathcal{M}_{s,p} \models \langle s \rangle p \text{ iff } \mathcal{M}_{s,p} \models [s]\neg p.$$
$$\overline{\mathcal{M}}_s, I(s) \not\models p \text{ iff } \underline{\mathcal{M}}_s, I(s) \models \neg p.$$

10.4.4 Categorical Syllogism

Categorical syllogism is a syllogism consisting of exactly three categorical propositions (two premises and a conclusion). By *categorical proposition*, we mean a proposition in which the predicate is affirmed or denied of all or part of the subject, There are four categorical propositions in Aristotle's syllogism, i.e. **A, E, I** and **O**.

Here, we discuss categorical syllogism in our approach. In contrast with Sect. 10.4.2, here we take a top-down approach, i.e., without describing the details of an underlying model $\mathcal{M}$, we simply assume the existence of such basic model so that we can perform reasoning process.

For example, when we are given a universal sentence 'All s is p', we at once construct a model $\mathcal{M}_{s,p}$ (or their lower model) and we assume it is a result of zooming in of $\mathcal{M}$ through $\{s, p\}$ for some $\mathcal{M}$.

There are four basic patterns of syllogism in Aristotle's syllogism such as BARBARA, CELARENT, DARII, and FARIO. Here we illustrate the inference process of the first pattern in our setting.

The form of BARBARA and its translation are given by

BARBARA	(Zooming in)	(Lower zooming in)
All m is p.	$M_{m,p} \models [m]p$ $\underline{M_m}, I(m) \models p$	$I(m) \subseteq \underline{R^m}(I(p))$
All s is m.	$M_{s,m} \models [s]m$ $\underline{M_s}, I(s) \models m$	$I(s) \subseteq R^s(I(m))$
All s is p.	$M_{s,p} \models [s]p$ $\underline{M_s}, I(s) \models p$	$I(s) \subseteq \underline{R^s}(I(p))$

First we describe the (simple) zooming case. By the premises we can assume the following two models:

$M^1_{m,p}$		m p...		$M^2_{s,m}$		s m...
$U_{m,p}$	$I(m) \cap I(p)$	1 1		$U_{s,m}$	$I(s) \cap I(m)$	1 1
	$I(m) \cap I(p)^c$	(discarded)			$I(s) \cap I(m)^c$	(discarded)
	$I(m)^c \cap I(p)$	0 1			$I(s)^c \cap I(m)$	0 1
	$I(m)^c \cap I(p)^c$	0 0			$I(s)^c \cap I(m)^c$	0 0

where the second rows in each valuation are discarded because $I(m) \cap I(p)^c = \emptyset$ and $I(s) \cap I(m)^c = \emptyset$ (we can assume they do not exist by the premise). To merge the two models, we make a zooming out of each model through $\{s, m, p\}$.

		$M^1_{s,m,p} \stackrel{\text{def}}{=} O^{m,p}_{s,m,p}(M^1_{m,p})$	$M^2_{s,m,p} \stackrel{\text{def}}{=} O^{s,m}_{s,m,p}(M^2_{s,m})$
		s m p...	s m p...
$U_{s,m,p}$	$I(s) \cap I(m) \cap I(p)$	1 1 1	1 1 1
	$I(s) \cap I(m) \cap I(p)^c$	(discarded)	1 1 0
	$I(s) \cap I(m)^c \cap I(p)$	1 1 0	(discarded)
	$I(s) \cap I(m)^c \cap I(p)^c$	1 0 0	(discarded)
	$I(s)^c \cap I(m) \cap I(p)$	0 1 1	0 1 1
	$I(s)^c \cap I(m) \cap I(p)^c$	(discarded)	0 0 1
	$I(s)^c \cap I(m)^c \cap I(p)$	0 0 1	0 0 1
	$I(s)^c \cap I(m)^c \cap I(p)^c$	0 0 0	0 0 0

By merging them, we have:

$M^1_{e,m,p} \circ M^2_{s,m,p}$		s m p
$U_{s,m,p}$	$I(s) \cap I(m) \cap I(p)$	1 1 1
	$I(s)^c \cap I(m) \cap I(p)$	0 1 1
	$I(s)^c \cap I(m)^c \cap I(p)$	0 0 1
	$I(s)^c \cap I(m)^c \cap I(p)^c$	0 0 0

to which, we again apply zooming in from $\{s, m, p\}$ to $\{s, p\}$:

$\mathcal{M}^1_{s,m,p} \circ \mathcal{M}^2_{s,m,p}$		s	m	p	$\beth^{s,m,p}_{s,p}$	$\mathcal{M}^3_{s,p}$		s	p
$U_{s,m,p}$	$I(s) \cap I(m) \cap I(p)$	1	1	1	$\longrightarrow$	$U_{s,p}$	$I(s) \cap I(p)$	1	1
	$I(s)^c \cap I(m) \cap I(p)$	0	1	1	Zooming in		$I(s)^c \cap I(p)$	0	1
	$I(s)^c \cap I(m)^c \cap I(p)$	0	0	1			$I(s)^c \cap I(p)^c$	0	0
	$I(s)^c \cap I(m)^c \cap I(p)^c$	0	-	0	0				

where $\mathcal{M}^3_{s,p} = I^{s,m,p}_{s,p}(\mathcal{M}^1_{s,m,p} \circ \mathcal{M}^2_{s,m,p})$. Thus, we have:

$$\mathcal{M}^3_{s,p} \models [s]p \text{ (and thus } \underline{\mathcal{M}^3_s}, I(s) \models p).$$

Hence, the process of BARBARA is performed on the basis of the following combination of zoomig and out:

$$\mathcal{M}^3_{s,p} = I^{s,m,p}_{s,p}(O^{m,p}_{s,m,p}(\mathcal{M}^1_{m,p}) \circ O^{s,m}_{s,m,p}(\mathcal{M}^2_{s,m}))).$$

Others can be similarly described.

CELARENT	(Zooming in)	(Lower zooming in)	
No m is p.	$M \models [m]\neg p$	$\mathcal{B}, m(M), I(m) \models \neg p$	$I(m) \subseteq \underline{R^m}(I(p)^c)$
All s is m.	$M \models [s]m$	$\mathcal{B}, s(M), I(s) \models m$	$I(s) \subseteq \underline{R^s}(I(m))$
No s is p.	$M \models [s]\neg p$	$\mathcal{B}, s(M), I(s) \models \neg p$	$I(s) \subseteq \underline{R^s}(I(p)^c)$

DARII	(Bottom-up)	(Top-down)	
All m is p.	$M \models [m]p$	$\mathcal{B}, m(M), I(m) \models p$	$I(m) \subseteq \underline{R^m}(I(p))$
Some s is m.	$M \models \langle s \rangle m$	$\mathcal{B}, s(M), I(s) \models m$	$I(s) \subseteq \overline{R^s}(I(m))$
Some s is p.	$M \models \langle s \rangle p$	$\mathcal{B}, s(M), I(s) \models p$	$I(s) \subseteq \overline{R^s}(I(p))$

FERIO	(Bottom-up)	(Top-down)	
No m is p.	$M \models [m]\neg p$	$\mathcal{B}, m(M), I(m) \models \neg p$	$I(m) \subseteq \underline{R^m}(I(p)^c)$
Some s is m.	$M \models \langle s \rangle m$	$\mathcal{B}, s(M), I(s) \models m$	$I(s) \subseteq \overline{R^s}(I(m))$
Some s is not p.	$M \models \langle s \rangle \neg p$	$\mathcal{B}, s(M), I(s) \models p$	$I(s) \subseteq \overline{R^s}(I(p)^c)$

10.5 Conclusions

In this chapter, we introduced the two operations of 'zooming in and out' as representing one aspect of granular computing in a logical setting and then applied them into a formulation of Aristotle's syllogism. We discuss universal and particular sentences, conversion and categorical syllogism in connection with 'zooming in and out'.

The subject discussed in this chapter thus reveals one of the rough set-based approaches to reasoning. This is because Aristotle's syllogism is known as a fundamental theory of our ordinary reasoning. Our approach also can give a model for Aristotle's syllogism based on Kripke semantics for modal logic.

It is an interesting topic to extend the present approach for predicative reasoning and common-sense reasoning processes. Other applications of 'zooming in and out' to reasoning may be found in Akama et al. [1].

References

1. Akama, S., Murai, T., Kudo, Y.: Reasoning with Rough Sets. Springer, Heidelberg (2018)
2. Chellas, B.: Modal Logic: An Introduction. Cambridge University Press, Cambridge (1980)
3. Lin, T.Y.: Granular computing on binary relation, I data mining and neighborhood systems, II rough set representations and belief functions. In: L. Polkowski, A. Skowron (eds.) Rough Sets in Knowledge Discovery 1: Methodology and Applications, pp. 107–121, 122–140. Physica-Verlag, Heidelberg (1998)
4. Murai, T., Nakata, M., Sato, Y.: A note on filtration and granular reasoning. Terano, T. et al. (eds.) New Frontiers in Artificial Intelligence, LNAI 2253, pp. 385–389. Springer (2001)
5. Murai, T., Resconi, G., Nakata, M., Sato, Y.: Operations of zooming in and out on possible worlds for semantic fields. In: Damiani, E. et al. (eds.) Knowledge-Based Intelligent Information Engineering Systems and Allied Technologies, pp. 1083–1087 (2002)
6. Murai, T., Resconi, G., Nakata, M., Sato, Y.: Granular reasoning using zooming in & out: Part 1. Propositional Reasoning. In: Proceedings of International Conference on Rough Sets, Fuzzy Sets, Data Mining, and Granular Computing (2003)
7. Pawlak, P.: Rough sets. Int. J. Comput. Inf. Sci. **11**, 341–356 (1982)
8. Pawlak, P.: Rough Sets: Theoretical Aspects of Reasoning about Data. Kluwer, Dordrecht (1991)
9. Resconi, G., Murai, T., Shimbo, M.: Field theory and modal logic by semantic field to make uncertainty emerge from information. Int. J. Gen. Syst. **29**, 737–782 (2000)
10. Skowron, A.: Toward intelligent systems: calculi of information granules. In: Terano et al. (eds.) New Frontiers in Artificial Intelligence, LNAI 2253, pp. 251–260, Springer (2001)

Chapter 11
A Modal Characterization of Visibility and Focus in Granular Reasoning

Abstract This chapter proposes two key concepts-focus and visibility-as modalities of modal logic. Scott-Montague models that we have proposed represent properties of visibility and focus and the concept that p is visible as modal sentence Vp and p is clearly visible-or is in focus-as modal sentence Cp.

Keywords Granular reasoning · Modal logic · Scott-montague model · Visibility · Focus

11.1 Introduction

Granular computing, based on rough set theory [12, 13], provides the basis for a new computing paradigm [7, 15]. Applying granular computing to logical reasoning processes, we have proposed granular reasoning for connecting possible world semantics and granular computing [8], and developed a granular reasoning framework called a *zooming reasoning system* [9–11].

Visibility separates all sentences into visible sentences, i.e., sentences we consider, and invisible sentences, which we do not consider. We have also constructed four-valued truth valuations based onvisibility and focus, which illustrate the concepts of clearly visible, obscurely visible, and invisible [4, 5]. We have not, however, considered other types of visibility and focus, e.g., by modality or algebraic structures.

The key concept of zooming reasoning system is focus that represents atomic sentences that appear in each step of reasoning. It is enough to consider truth values of atomic sentences used in the current step, letting us ignore truth values of other atomic sentences not used by constructing equivalent classes of possible worlds based on truth values of the atomic sentences appearing in the current reasoning step.

Focus thus provides a three-valued truth valuation that assigns truth values true or false to atomic sentences that appear in focus and assigns a truth value unknown to other atomic sentences. In the zooming reasoning system, controlling the size of equivalent classes of possible worlds corresponds to reasoning.

© Springer Nature Switzerland AG 2020
S. Akama et al., *Topics in Rough Set Theory*, Intelligent Systems
Reference Library 168, https://doi.org/10.1007/978-3-030-29566-0_11

In redefining the concept, we have introduced the granularity concept of visibility [4]. In granular reasoning, visibility and focus are analogies of terms in vision, representing the following concepts:

(1) Visibility: the set of atomic sentences that must be considered in the current step of reasoning, corresponding to focus in the zooming reasoning system.

(2) Focus (redefined): the set of atomic sentences for which truth values are decidable as either true or false.

Visibility capture the concepts of visibility and focus as modalities, producing Scott-Montague models that represent properties of visibility and focus and represent the concept that p is visible as modal sentence Vp and p is clearly visible-or is in focus-as modal sentence Cp. Note that this is a revised, extended version of our conference paper [6].

11.2   Background

11.2.1   Modal Logic, Kripke and Scott-Montague Models

Let $\mathcal{P}$ be a set of (at most countably infinite) atomic sentences. We construct language $\mathcal{L}_{ML}(\mathcal{P})$ for modal logic from $\mathcal{P}$ using logical operators $\top$ (truth constant), $\bot$ (falsity constant), $\neg$ (negation), $\wedge$ (conjunction), $\vee$ (disjunction), $\rightarrow$ (material implication), $\leftrightarrow$ (equivalence), and two modal operators $\Box$ (necessity) and $\Diamond$ (possibility) and using the following rules:

(1) $p \in \mathcal{P} \Rightarrow p \in \mathcal{L}_{ML}(\mathcal{P})$, (2) $p \in \mathcal{L}_{ML}(\mathcal{P}) \Rightarrow \neg p \in \mathcal{L}_{ML}(\mathcal{P})$, (3) $p, q \in \mathcal{L}_{ML}(\mathcal{P}) \Rightarrow p \wedge q, p \vee q, p \rightarrow q, p \leftrightarrow q \in \mathcal{L}_{ML}(\mathcal{P})$, (4) $p \in \mathcal{L}_{ML}(\mathcal{P}) \Rightarrow \Box p, \Diamond p \in \mathcal{L}_{ML}(\mathcal{P})$. A sentence is called *non-modal* if it contains no modal operators. We denote $\mathcal{L}(\mathcal{P})$ to mean the set of all non-modal sentences.

Kripke models [1] are semantic frameworks that provide possible world semantics for modal logic. A Kripke model is a triple $M = \langle W, R, v \rangle$, where W is a non-empty set of possible worlds, R is a binary relation on W called an *accessibility relation*, and v is a truth valuation function that assigns truth value $\mathbf{t}$ (true) or $\mathbf{f}$ (false) to each atomic sentence $p \in \mathcal{P}$ in each world $w \in W$.

We denote $M, w \models p$ to mean that sentence p is true at possible world w in model M. Relationship $M, w \models p$ is obtained by extending valuation in the usual way. For any sentence $p \in \mathcal{L}_{ML}(\mathcal{P})$, we define the set of p in M as $\|p\|^{M} = \{w \in W \mid M, w \models p\}$. Truth conditions of modal sentences in Kripke model M are given by

$$M, w \models \Box p \Leftrightarrow \forall x \in W(wRx \Rightarrow M, x \models p),$$
$$M, w \models \Diamond p \Leftrightarrow \exists x \in W(wRx \text{ and } M, x \models p).$$

Conditions of accessibility relation R such that reflexivity, symmetry, and transitivity correspond to axiom schemata of modal logic.

Scott-Montague models-or minimal models [1]-are a generalization of Kripke models. A Scott-Montague model is a triple

$$M = \langle W, N, v \rangle,$$

where W is a non-empty set of possible worlds and v is a truth valuation function used in Kripke models. Instead of accessibility relation R, the Scott-Montague model uses function N from W to 2^{2^W}. Truth conditions of modal sentences in the Scott-Montague model M are given by

$$M, w \models \Box p \Leftrightarrow \|p\|^M \in N(w).$$
$$M, w \models \Diamond p \Leftrightarrow (\|p\|^M)^c \notin N(w).$$

Conditions of N are considered such that

(m) $X \cap Y \in N(w) \Rightarrow X \in N(w)$ and $Y \in N(w)$,
(c) $X, Y \in N(w) \Rightarrow X \cap Y \in N(w)$,
(n) $W \in N(w)$,
(d) $X \in N(w) \Rightarrow X^c \notin N(w)$,
(t) $X \in N(w) \Rightarrow w \in X$,
(4) $X \in N(w) \Rightarrow \{x \in W \mid X \in N(x)\} \in N(w)$,
(5) $X \notin N(w) \Rightarrow \{x \in W \mid X \notin N(x)\} \in N(w)$.

For every Kripke model $M_K = \langle W, R, v \rangle$ there is equivalent Scott-Montague model $M_{SM} = \langle W, N, v \rangle$ such that $M_K, w \models p$ iff $M_{SM}, w \models p$ for any sentence $p \in \mathcal{L}_{ML}(\mathcal{P})$ and any possible world $w \in W$. From accessibility relation R in Kripke model M, function N in Scott-Montague model M_{SM} is constructed by

$$x \in N(w) \Leftrightarrow \{x \in W \mid wRx\}$$

for every w and every $X \subseteq W$. Function N in M_{SM} satisfies above conditions (m), (c), and (n), and the following condition:

(a) $X \in N(w) \Leftrightarrow \bigcup N(w) \subseteq X$,

where $\bigcup N(w)$ is the intersection of all sets in $N(w)$. Kripke models are thus special cases of Scott-Montague models.

Smallest classical modal logic E is proved to be both sound and complete with respect to the class of all Scott-Montague models, where E contains schema $D\Diamond$: $\Diamond p \leftrightarrow \neg \Box \neg p$ and rule of inference

RE: from $p \leftrightarrow q$ infer $\Box p \leftrightarrow \Box q$

with rules and axiom schemata of propositional logic. Each condition of N corresponds to axiom schema such that

(M): $\Box(p \wedge q) \rightarrow (\Box p \wedge \Box q)$,
(C): $(\Box p \wedge \Box q) \rightarrow \Box(p \wedge q)$,
(N): $\Box \top$,
(D): $\Box p \rightarrow \Diamond p$,
(T): $\Box p \rightarrow p$,
(4): $\Box p \rightarrow \Box \Box p$,
(5): $\Diamond p \rightarrow \Box \Diamond p$.

11.3 Visibility and Focus: Two Granular Reasoning Concepts

11.3.1 Visibility: Separation of Visible and Invisible Sentences

Suppose we fix Kripke model $M = \langle W, ..., v \rangle$ with set of possible worlds W and truth valuation v, and consider representing granular reasoning based on model M. Note that we omit the accessibility relation in the Kripke model because we do not use the accessibility relation between possible worlds here.

Let $\Gamma \subset \mathcal{L}(\mathcal{P})$ be a finite set of non-modal sentences considered in the current reasoning step. Formally, definitios of visibility $Vs(\Gamma)$ and focus $Fc(\Gamma)$ relative to Γ are as follows:

$$Vs(\Gamma) = \mathcal{P} \cap \text{Sub}(\Gamma) = \mathcal{P}_\Gamma,$$
$$Fc(\Gamma) = \{p \in \mathcal{P} \mid \text{either } \Gamma \models p \text{ or } \Gamma \models \neg p\},$$

where $\text{Sub}(\Gamma)$ is the set of all subsentences of sentences in Γ and $\Gamma \models p$ means that, for any possible world $w \in W$, if $M, w \models q$ for any sentence $q \in \Gamma$ then $M, w \models p$. Using truth valuation function v, we construct agreement relation $R_{V_s(\Gamma)}$ based on visibility $Vs(\Gamma)$ as follows.

$$x R_{V_s(\Gamma)} y \Leftrightarrow v(p, x) = v(p, y), \forall p \in Vs(\Gamma).$$

Agreement relation $R_{V_s(\Gamma)}$ is easily confirmed as an equivalence relation, and $R_{V_s(\Gamma)}$ therefore induces quotient set $\tilde{W} = W / R_{V_s(\Gamma)}$ of W, i.e., the set of all equivalent classes of possible worlds in W. We also construct truth valuation $\tilde{v}_{V_s(\Gamma)}$ for equivalent classes of possible worlds $\tilde{x} = [x]_{R_{V_s(\Gamma)}} \in \tilde{W}$. Valuation $\tilde{v}_{V_s(\Gamma)}$ becomes the following three-valued:

$$\tilde{v}_{V_s(\Gamma)} : \mathcal{P} \times \tilde{W} \to 2^{\{t,f\}} \setminus \{\{t, f\}\}.$$

Three-valued valuation $\tilde{v}_{V_s(\Gamma)}$ is defined by

$$\tilde{v}_{V_s(\Gamma)}(p, \tilde{w}) = \begin{cases} \{t\}, & \text{if } v(p, x) = t, \forall x \in \tilde{w}, \\ \{f\}, & \text{if } v(p, x) = f, \forall x \in \tilde{w}, \\ \emptyset, & \text{otherwise.} \end{cases}$$

Below, we use the following notations: $\mathbf{T} = \{t\}$ and $\mathbf{F} = \{f\}$. Using $\tilde{v}_{V_s(\Gamma)}$, we define the visibility of atomic sentences.

Definition 11.1 Atomic sentence p is *visible* at $\tilde{w}$ iff either $\tilde{v}_{V_s(\Gamma)}(p, \tilde{w}) = \mathbf{T}$ or $\tilde{v}_{V_s(\Gamma)}(p, \tilde{w}) = \mathbf{F}$. p is *invisible* at $\tilde{w}$ iff $\tilde{v}_{V_s(\Gamma)}(p, \tilde{w}) = \emptyset$.

Three-valued valuation $\tilde{v}_{V_s(\Gamma)}$ is extended to any non-modal sentence by truth assignment of connectives $\neg$ (negation), $\wedge$ (conjunction), $\vee$ (disjunction), and $\to$ (implication) as in Table 11.1.

Table 11.1 Truth tables of the three-valued valuation

Negation $\neg p$				Conjunction $p \wedge q$			
	p	$\neg p$		$p \backslash q$	$\emptyset$	**F**	**T**
	$\emptyset$	$\emptyset$		$\emptyset$	$\emptyset$	$\emptyset$	$\emptyset$
	F	**T**		**F**	$\emptyset$	**F**	**F**
	T	**F**		**T**	$\emptyset$	**F**	**T**
Disjunction $p \vee q$				Implication $p \rightarrow q$			
$p \backslash q$	$\emptyset$	**F**	**T**	$p \backslash q$	$\emptyset$	**F**	**T**
$\emptyset$	$\emptyset$	$\emptyset$	$\emptyset$	$\emptyset$	$\emptyset$	$\emptyset$	$\emptyset$
F	$\emptyset$	**F**	**T**	**F**	$\emptyset$	**T**	**T**
T	$\emptyset$	**F**	**T**	**T**	$\emptyset$	**F**	**T**

We denote extended three-valued truth valuation by the same notation $\tilde{v}_{V_s(\Gamma)}$. As with atomic sentences, for any non-modal sentence p. we say p is visible at $\tilde{w}$ iff either $\tilde{v}_{V_s(\Gamma)}(p, \tilde{w}) = \mathbf{T}$ or $\tilde{v}_{V_s(\Gamma)}(p, \tilde{w}) = \mathbf{F}$. p is invisible at $\tilde{w}$ iff $\tilde{v}_{V_s(\Gamma)}(p, \tilde{w}) = \emptyset$.

If both p and q are visible, $\neg p$, $p \wedge q$, $p \vee q$, and $p \rightarrow q$ are also visible. If p is invisible, then $\neg p$ is also invisible. If either p or q is invisible, then $p \wedge q$, $p \vee q$, and $p \rightarrow q$ are also invisible.

As stated in Sect. 11.1, we intend for the concept of visibility to separate all sentences into visible sentences. i.e., sentences we consider, and invisibles entences which we do not consider. In the definition of three-valued truth valuation, we intend that the interpretation of truth value $\emptyset$ is "not considered", and letting the truth value of all invisible sentences be $\emptyset$ via three-valued truth tables (Table 11.1), we separate all non-modal sentences into visible and invisible sentences by truth values because visibility $V_s(\Gamma)$ consists of atomic sentences considered in the current reasoning step, meaning that any other sentences that contain some invisible atomic sentence $p \notin V_s(\Gamma)$ are also invisible, that is, not considered in the current reasoning step. For extending the concept of visibility naturally to every non-modal sentence, we consider that the truth value of such invisible sentences should be neither $\mathbf{T}$ nor $\mathbf{F}$, and therefore should be $\emptyset$.

11.3.2 Focus: Separation of Clearly Visible and Obscurely Visible Sentences

Using focus $F_c(\Gamma)$ relative to Γ, we construct agreement relation $R_{F_c(\Gamma)}$ on quotient set $\tilde{W}$ of possible worlds. If $F_c(\Gamma) \neq \emptyset$, we define agreement relation $R_{F_c(\Gamma)}$ as follows:

$$\tilde{x} R_{F_c(\Gamma)} \tilde{y} \Leftrightarrow \tilde{v}_{V_s(\Gamma)}(p, \tilde{x}) = \tilde{v}_{V_s(\Gamma)}(p, \tilde{y}), \forall p \in F_c(\Gamma).$$

Agreement relation $R_{F_c(\Gamma)}$ on $\tilde{W}$ induces quotient set $\hat{W} = \tilde{W}/R_{F_c(\Gamma)}$. We treat each equivalent class $\hat{W} = [\tilde{w}]_{R_{F_c(\Gamma)}}$ as a unit of consideration as if each $\hat{w}$ were a "possible world'. If $F_c(\Gamma) = \emptyset$, we cannot construct the agreement relation, so, we define $\hat{W} = \{\tilde{W}\}$.

We consider valuation function $\hat{v}_{F_c(\Gamma)}$ for equivalent classes in $\hat{W}$ as the following four-valued:

$$\hat{v}_{F_c(\Gamma)} : \mathcal{P} \times \hat{W} \to 2^{\{\mathbf{T},\mathbf{F}\}}.$$

Valuation $\hat{v}_{F_c(\Gamma)}$ is defined by:

$$\hat{v}_{F_c(\Gamma)}(p, \hat{w}) = \begin{cases} \{\mathbf{T}\}, & \tilde{v}_{V_s(\Gamma)}(p, \tilde{x}) = \mathbf{T}, \forall \tilde{x} \in \hat{w}, \\ \{\mathbf{F}\}, & \tilde{v}_{V_s(\Gamma)}(p, \tilde{x}) = \mathbf{F}, \forall \tilde{x} \in \hat{w}, \\ \{\mathbf{T},\mathbf{F}\}, & \exists \tilde{x}, \tilde{y} \in \hat{w} \text{ such that } \tilde{v}_{V_s(\Gamma)}(p, \tilde{x}) = \mathbf{T}, \text{ and } \tilde{v}_{V_s(\Gamma)}(p, \tilde{y}) = \mathbf{F}, \\ \emptyset, & \text{otherwise.} \end{cases}$$

Definition 11.2 Atomic sentence p is *clearly visible*-or *in foucs*-at $\hat{w}$ iff either $\hat{v}_{F_c(\Gamma)}(p, \hat{w}) = \{\mathbf{T}\}$ or $\hat{v}_{F_c(\Gamma)}(p, \hat{w}) = \{\mathbf{F}\}$. p is *obscurely visible* at $\hat{w}$ iff $\hat{v}_{F_c(\Gamma)}(p, \hat{w}) = \{\mathbf{T},\mathbf{F}\}$. p is *invisible* at $\hat{w}$ iff $\hat{v}_{F_c(\Gamma)}(p, \hat{w}) = \emptyset$.

This definition makes it clear that, for all $p \in F_c(\Gamma)$, p is clearly visible at all $\hat{w} \in \hat{W}$.

As with three-valued valuation $\tilde{v}_{V_s(\Gamma)}$, four-valued valuation $\hat{v}_{V_s(\Gamma)}$ is extended to any non-modal sentences by truth assignment of connectives (Table 11.2).

We denote extende four-valued truth valuation using same notation $\hat{v}_{F_c(\Gamma)}$. As with the three-valued case, for any clearly visible sentences p and q, $\neg p$, $p \wedge q$, $p \vee q$ and $p \to q$ are also clearly visible. There is thus at least one equivalence class $\hat{w} \in \hat{W}$ such that $\hat{v}_{F_c(\Gamma)}(p, \hat{w}) = \{\mathbf{T}\}$ for all $p \in \Gamma$.

Table 11.2 Truth tables of the four-valued valuation

Negation $\neg p$				Disjunction $p \vee q$				
		p	$\neg p$		$p\backslash q$	$\emptyset$	$\{\mathbf{F}\}$	$\{\mathbf{T}\}$	$\{\mathbf{T},\mathbf{F}\}$
		$\emptyset$	$\emptyset$		$\emptyset$	$\emptyset$	$\emptyset$	$\emptyset$	$\emptyset$
		$\{\mathbf{F}\}$	$\{\mathbf{T}\}$		$\{\mathbf{F}\}$	$\emptyset$	$\{\mathbf{F}\}$	$\{\mathbf{T}\}$	$\{\mathbf{T},\mathbf{F}\}$
		$\{\mathbf{T}\}$	$\{\mathbf{F}\}$		$\{\mathbf{T}\}$	$\emptyset$	$\{\mathbf{T}\}$	$\{\mathbf{T}\}$	$\{\mathbf{T}\}$
		$\{\mathbf{T},\mathbf{F}\}$	$\{\mathbf{T},\mathbf{F}\}$		$\{\mathbf{T},\mathbf{F}\}$	$\emptyset$	$\{\mathbf{T},\mathbf{F}\}$	$\{\mathbf{T}\}$	$\{\mathbf{T},\mathbf{F}\}$

Conjunction $p \wedge q$					Implication $p \to q$				
$p\backslash q$	$\emptyset$	$\{\mathbf{F}\}$	$\{\mathbf{T}\}$	$\{\mathbf{T},\mathbf{F}\}$	$p\backslash q$	$\emptyset$	$\{\mathbf{F}\}$	$\{\mathbf{T}\}$	$\{\mathbf{T},\mathbf{F}\}$
$\emptyset$	$\emptyset$	$\emptyset$	$\emptyset$	$\emptyset$	$\emptyset$	$\emptyset$	$\emptyset$	$\emptyset$	$\emptyset$
$\{\mathbf{F}\}$	$\emptyset$	$\{\mathbf{F}\}$	$\{\mathbf{F}\}$	$\{\mathbf{F}\}$	$\{\mathbf{F}\}$	$\emptyset$	$\{\mathbf{T}\}$	$\{\mathbf{T}\}$	$\{\mathbf{T}\}$
$\{\mathbf{T}\}$	$\emptyset$	$\{\mathbf{F}\}$	$\{\mathbf{T}\}$	$\{\mathbf{T},\mathbf{F}\}$	$\{\mathbf{T}\}$	$\emptyset$	$\{\mathbf{F}\}$	$\{\mathbf{T}\}$	$\{\mathbf{T},\mathbf{F}\}$
$\{\mathbf{T},\mathbf{F}\}$	$\emptyset$	$\{\mathbf{F}\}$	$\{\mathbf{T},\mathbf{F}\}$	$\{\mathbf{T},\mathbf{F}\}$	$\{\mathbf{T},\mathbf{F}\}$	$\emptyset$	$\{\mathbf{T},\mathbf{F}\}$	$\{\mathbf{T}\}$	$\{\mathbf{T},\mathbf{F}\}$

Note, however, that not all two-valued tautologies are satisfied by $\hat{v}_{F_c(\Gamma)}$. For any invisible sentence p and obscurely visible sentence q, for example, excluded middle is not satisfied: $\hat{v}_{F_c(\Gamma)}(p \vee \neg p, \hat{w}) = \emptyset$ and $\hat{v}_{F_c(\Gamma)}(q \vee \neg q, \hat{w}) = \{T, F\}$ for all $\hat{w} \in \hat{W}$.

In the definition of four-valued valuation, we intend that truth values $\{T\}$ and $\{F\}$ correspond to true and false in the sense of two-valued truth valuation. The interpretation of truth value $\{T, F\}$ we intend is "unknown" and $\emptyset$ is "not considered".

As defined in Definition 11.2, we intend that the concept of focus separates all visible sentences into clearly visible sentences, i.e., sentences whose truth values are uniquely determined to be either true or false in the current reasoning step, and obscurely visible sentences that appear in the current reasoning step, but those truth values are not uniquely determined. With this intention, we set the truth value of obscurely visible atomic sentences by $\{T, F\}$.

We also consider that all invisible sentences under the concept of visibility should still be invisible even though the concept of focus is applied. Based on the definition of three-valued truth valuation, we set the truth value of invisible atomic sentences by $\emptyset$.

To extend the concept of focus naturally to every nonmodal sentence as with visibility, we set the truth value of any nonmodal sentence (Table 11.2). Given that truth values of sentences $p, q \in \mathcal{L}(\mathcal{P})$ are either $\{T\}$ or $\{F\}$ or $\{T, F\}$, truth valuations of sentences $\neg p$, $p \vee q$, $p \wedge q$, and $p \rightarrow q$ (Table 11.2) are identical to Kleene's strong three-valued logic [3]. Based on the above discussion on invisible sentences, we set the truth value of sentences that contain invisible sentences by $\emptyset$ (Table 11.2).

Example 11.1 Let $\mathcal{P} = \{p, q, r\}$ be a set of atomic sentences and W be a non-empty set that has the following eight possible worlds.

$$w_1 = \{p, q, r\}, \ w_2 = \{p, q\}, \ w_3 = \{p, r\}, \ w_4 = \{p\},$$
$$w_5 = \{q, r\}, \quad w_6 = \{q\}, \quad w_7 = \{r\}, \quad w_8 = \emptyset.$$

We define the truth value of each atomic sentence $p \in \mathcal{P}$ at each world $w \in W$ by $v(p, w) = t \Leftrightarrow p \in w$, With this truth assignment, for example, all atomic sentences are true at w_1 but all atomic sentences are false at w_8.

Suppose we have the following set of nonmodal sentences considered in the current reasoning step $\Gamma = \{q, p \rightarrow q\}$. We have visibility $V_s(\Gamma)$ and focus $F_c(\Gamma)$ relative to Γ as follows: $V_s(\Gamma) = \{p, q\}$, $F_c((\Gamma) = \{q\}$.

Constructing agreement relation $R_{V_s(\Gamma)}$, we have the following equivalent classes of possible worlds:

$$\tilde{w}_1 = \{w_1, w_2\}, \ \tilde{w}_3 = \{w_3, w_4\},$$
$$\tilde{w}_5 = \{w_5, w_6\}, \ \tilde{w}_7 = \{w_7, w_8\}.$$

Each atomic sentence has the following three-valued truth value:

$$\tilde{v}_{F_c(\Gamma)}(p, \tilde{w}_1) = T, \ \tilde{v}_{F_c(\Gamma)}(q, \tilde{w}_1) = T, \ \tilde{v}_{F_c(\Gamma)}(r, \tilde{w}_1) = \emptyset,$$
$$\tilde{v}_{F_c(\Gamma)}(p, \tilde{w}_3) = T, \ \tilde{v}_{F_c(\Gamma)}(q, \tilde{w}_3) = F, \ \tilde{v}_{F_c(\Gamma)}(r, \tilde{w}_3) = \emptyset,$$
$$\tilde{v}_{F_c(\Gamma)}(p, \tilde{w}_5) = F, \ \tilde{v}_{F_c(\Gamma)}(q, \tilde{w}_5) = T, \ \tilde{v}_{F_c(\Gamma)}(r, \tilde{w}_5) = \emptyset,$$
$$\tilde{v}_{F_c(\Gamma)}(p, \tilde{w}_7) = F, \ \tilde{v}_{F_c(\Gamma)}(q, \tilde{w}_7) = F, \ \tilde{v}_{F_c(\Gamma)}(r, \tilde{w}_7) = \emptyset.$$

These truth values indicate that p and q are visible, while r is invisible.

We next construct agreement relation $R_{F_c(\Gamma)}$ on $\tilde{W}$ and get the following two equivalent classes:

$$\hat{w}_1 = \{\hat{w}_1, \hat{w}_5\} = \{\{w_1, w_2\}, \{w_5, w_6\}\},$$
$$\hat{w}_3 = \{\hat{w}_3, \hat{w}_7\} = \{\{w_3, w_4\}, \{w_7, w_8\}\}.$$

Each atomic sentence has the following four-values truth value:

$$\hat{v}_{F_c(\Gamma)}(p, \hat{w}_1) = \{\mathbf{T}, \mathbf{F}\}, \ \hat{v}_{F_c(\Gamma)}(q, \hat{w}_1) = \{\mathbf{T}\}, \ \hat{v}_{F_c(\Gamma)}(r, \hat{w}_1) = \emptyset,$$
$$\hat{v}_{F_c(\Gamma)}(p, \hat{w}_3) = \{\mathbf{T}, \mathbf{F}\}, \ \hat{v}_{F_c(\Gamma)}(q, \hat{w}_3) = \{\mathbf{F}\}, \ \hat{v}_{F_c(\Gamma)}(r, \hat{w}_3) = \emptyset.$$

This means that q is clearly visible and p is obscurely visible. Similar to the three-valued case, r is invisible. Four-valued truth values of any non-modal sentences are calculated based on Table 11.2.

For example, the truth value of $p \rightarrow q$ is: $\hat{v}_{F_c(\Gamma)}(p \rightarrow q, \hat{w}_1) = \{\mathbf{T}\}$ and $\hat{v}_{F_c(\Gamma)}(p \rightarrow q, \hat{w}_3) = \{\mathbf{T}, \mathbf{F}\}$. All non-modal sentences in Γ are therefore true, i.e., clearly visible, at $\hat{w}_1$.

11.4   Granular Reasoning and Scott-Montague Models

11.4.1   Visibility as Modality

In considering the concept of visibility by modality based on Scott-Montague models, suppose we have quotient set $\tilde{W}$ of W by agreement relation $R_{V_s(\Gamma)}$ based on visibility $V_s(\Gamma)$ relative to Γ and non-modal sentence p is visible at $\tilde{w} \in \tilde{W}$.

Instead of modal operator $\square$, we use modal operator V, and we read Vp as p is visible. We intend to represent visibility by Scott-Montague model $\mathcal{M}_{V_s(\Gamma)}$ as follows for each possible world $x \in \tilde{w}$:

$$\mathcal{M}_{V_s(\Gamma)}, x \models Vp, \text{ if } p \text{ is visible at } \tilde{w}.$$

To represent the concept of visibility as modality, we use the following simple function $N_{V_s(\Gamma)}$:

Definition 11.3 Let $\tilde{W} = \{\tilde{w}_1, ..., \tilde{w}_n\}$ be the equivalent classes of possible worlds based on $V_s(\Gamma)$. Function $N_{V_s(\Gamma)} : W \rightarrow 2^{2^W}$ is defined by

$$N_{V_s(\Gamma)}(x) = \{\bigcup A \mid A \subseteq \tilde{W}\}, \forall x \in W$$

where $\bigcup A$ means the union of all equivalent classes in A. If $A = \emptyset$, we define $\bigcup A = \emptyset$.

This means that, for any $x \in W$, each element $X \in N_{V_s(\Gamma)}(x)$ is constructed by the union of equivalent classes of possible worlds in $\tilde{W}$. Each $\tilde{w} \in \tilde{W}$ is equivalent class $[w]_{R_s(\Gamma)} \subseteq W$, so function $N_{V_s(\Gamma)}$ is well-defined. $N_{V_s(\Gamma)}$ satisfies the following conditions.

Lemma 11.1 *Constructed function $N_{V_s(\Gamma)}$ satisfies conditions (c), (n), (4), and (5). In addition, $N_{V_s(\Gamma)}$ satisfies the following property:*

$(v!)$ $X \in N_{V_s(\Gamma)}$ $\Leftrightarrow$ $X^c \in N_{V_s(\Gamma)}$.

Proof For (c), suppose $X \in N_{V_s(\Gamma)}(w)$ and $Y \in N_{V_s(\Gamma)}(w)$. By Definition 11.3, there are some $J, K \subseteq \tilde{W}$ such that $X = \bigcup J$ and $Y = \bigcup K$. It is easily confirmed that $X \cap Y = \bigcup \{\tilde{x} \cap \tilde{y} \mid \tilde{x} \in J, \tilde{y} \in K\}$, and either $\tilde{x} \cap \tilde{y} = \tilde{x}$ or $\tilde{x} \cap \tilde{y} = \emptyset$ because both $\tilde{x}$ and $\tilde{y}$ are equivalent classes of $\tilde{W}$. $X \cap Y$ is therefore a union of equivalent classes, which conclude $X \cap Y = N_{V_s(\Gamma)}(w)$.

For (n), it is trivial that $W = \bigcup \tilde{W} \in N_{V_s(\Gamma)}(w)$.

For (4), suppose $X \in N_{V_s(\Gamma)}(w)$. From Definition 11.3, we have $X \in N_{V_s(\Gamma)}(w)$ for all $w \in W$. Therefore $\{x \in W \mid X \in N(x)\} = W$ and $W \in N_{V_s(\Gamma)}(w)$ by (n).

For (5), suppose $X \notin N_{V_s(\Gamma)}(w)$. As with (4), we have $X \notin N_{V_s(\Gamma)}(w)$ for all $w \in W$ and therefore $\{x \in W \mid X \notin N(x)\} = W \in N_{V_s(\Gamma)}(w)$.

For (v!), suppose $X \in N_{V_s(\Gamma)}(w)$. There is some $J \subseteq \tilde{W}$ such that $X = \bigcup J$. Therefore $X = \bigcup (\tilde{W} \setminus J) \in N_{V_s(\Gamma)}$. The converse is similarly proved.

Note that function $N_{V_s(\Gamma)}$ does not satisfy property (m) in general and that because it satifies property (v!). It clearly does not satisfy properties (d) and (t).

Using $N_{V_s(\Gamma)}$, we construct Scott-Montague model $M_{V_s(\Gamma)} = \langle W, N_{V_s(\Gamma)}, v \rangle$ from fixed Kripke model M. Lemma 11.1 indicates that model $M_{V_s(\Gamma)}$ validates schmata C, N, 4, and 5. Condition (v!) corresponds to the following schma.

V! : $Vp \leftrightarrow V\neg p$.

The truth condition of modal sentences (5) captures part of the concept of visibility. The following lemma captures the property that if both p and q are visible, then $\neg p$, $p \wedge q$, $p \vee q$, and $p \rightarrow q$ are also visible:

Lemma 11.2 *1. If $p \in V_s(\Gamma)$, then $M_{V_s(\Gamma)}, w \models Vp$ for all $w \in W$.*
2. For any $p \in \mathcal{L}(\mathcal{P})$ and all $w \in W$, if $M_{V_s(\Gamma)}, w \models Vp$, then $M_{V_s(\Gamma)}, w \models V\neg p$.
3. For any $p, w \in V_s(\Gamma)$ and all $w \in W$, if both $M_{V_s(\Gamma)}, w \models Vp$ and $M_{V_s(\Gamma)}, w \models Vq$, then $M_{V_s(\Gamma)}, w \models V(p \wedge q)$, $M_{V_s(\Gamma)}, w \models V(p \vee q)$, and $M_{V_s(\Gamma)}, w \models V(p \rightarrow q)$.

Proof For property 1, suppose $p \in V_s(\Gamma)$ holds. By the definition of function $N_{V_s(\Gamma)}$, for every possible world $w \in W$, set $N_{V_s(\Gamma)}(w)$ consists of unions $\bigcup A$ of subsetes of equivalent class $A \subseteq \tilde{W}$ based on visibility $V_s(\Gamma)$. $\| p \| = \bigcup \{\tilde{x} \mid \tilde{v}_{V_s(\Gamma)}(p, \tilde{x} = \mathbf{T}\} \in N_{V_s(\Gamma)}(w)$ holds for every $w \in W$ by the definition of three-valued truth valuation $\tilde{v}_{V_s(\Gamma)}$.

Property 2 is clear from property (v!) in Lemma 11.1.

For property 3, suppose we have both $M_{V_s(\Gamma)}, w \models Vp$ and $M_{V_s(\Gamma)}, w \models Vq$. $M_{V_s(\Gamma)}, w \models V(p \wedge q)$ is clear from the assumption and the property of (c) in Lemma 11.1. For $M_{V_s(\Gamma)}, w \models V(p \vee q)$, it is trivial from the definition of $N_{V_s(\Gamma)}$. $M_{V_s(\Gamma)}, w \models V(p \rightarrow q)$ is also clear from $M_{V_s(\Gamma)}, w \models V(p \vee q)$.

Combining these lemmas yields the following result:

Theorem 11.1 *Let Γ be a non-empty set of non-modal sentences, $\tilde{W}$ be the set of equivalent classes of possible worlds based on the visibility $V_s(\Gamma)$, and $M_{V_s(\Gamma)} = \langle W, N_{V_s(\Gamma)}, v \rangle$ be a Scott-Montague model that has function $N_{V_s(\Gamma)}$. For any non-modal sentence $p \in \mathcal{L}(\mathcal{P})$, if p is visible at $\tilde{w} \in W$, then $M_{V_s(\Gamma)}, w \models Vp$ for all $x \in \tilde{w}$.*

Proof This is obvious from Lemmas 11.1 and 11.2.

The converse of Theorem 11.1 is not always satisfied because, in our assumption, any visible tautology p becomes $M_{V_s(\Gamma)}, w \models Vp$. Suppose, for example, that atomic sentence r is invisible by $\tilde{v}_{V_s(\Gamma)}$. Tautology $r \vee \neg r$ is also invisible by the definition of visibility.

Truth set $\|r \vee \neg r\|^{M_{V_s(\Gamma)}} = W$ is, however, an element of $N_{V_s(\Gamma)}(w)$ for all $w \in W$, therefore $M_{V_s(\Gamma)}, w \models V(r \vee \neg r)$. Unfortunately, we cannot avoid this difficulty because it causes the schema **N** to be satisfied using $N_{V_s(\Gamma)}$. We thus must restrict our formulation to satisfiable sentences.

Example 11.2 We use the same setting as for Example 11.1. Using $\tilde{W} = \{\tilde{w}_1, \tilde{w}_3, \tilde{w}_5, \tilde{w}_7\}$, we construct Scott-Montague model $M_{V_s(\Gamma)} = \langle W, N_{V_s(\Gamma)}, v \rangle$. Where W and v are the same set of possible worlds and the same turth valuation function defined in Example 11.1.

We have $V_s(\Gamma) = \{p, q\}$, so atomic sentences p and q are visible but r is invisible. For these atomic sentences, we have the following truth sets:

$$\|p\|^{M_{V_s(\Gamma)}} = \{w_1, w_2, w_3, w_4\}$$
$$\|q\|^{M_{V_s(\Gamma)}} = \{w_1, w_2, w_5, w_6\}$$
$$\|r\|^{M_{V_s(\Gamma)}} = \{w_1, w_3, w_5, w_7\}$$

It holds that $\|p\|^{M_{V_s(\Gamma)}} = \{w_1, w_2\} \cup \{w_3, w_4\} = \tilde{w}_1 \cup \tilde{w}_3$ and $\|q\|^{M_{V_s(\Gamma)}} = \{w_1, w_2\} \cup \{w_5, w_6\} = \tilde{w}_1 \cup \tilde{w}_5$. Therefore, for example, $M_{V_s(\Gamma)}, x \models Vp$ and $M_{V_s(\Gamma)}, x \models Vq$ for all $x \in \tilde{w}_1$ by truth condition (5). We cannot construct $\|r\|^{M_{V_s(\Gamma)}}$ by union of $\tilde{w}_i$, we have $M_{V_s(\Gamma)}, w \not\models Vr$ for all $w \in W$.

11.5 Focus as Modality

As with visibility, we capture the concept of focus by modality based on Scott-Montague models. Focus $F_c(\Gamma)$ relative to Γ divides all visible sentences into clearly visible and obscurely visible. For any clearly visible sentence p, using modal operator C, we denote Cp to mean that p is clearly visible.

We show the focus by some Scott-Montague model $\mathcal{M}_{F_c(\Gamma)}$ as follows: Let $\hat{W}$ be the quotient set of set $\tilde{W}$ by agreement relation $R_{F_c(\gamma)}$ based on focus $F_c(\Gamma)$, and $\hat{w} \in \hat{W}$ be an equivalent class of elements in $\hat{W}$. For each possible world $x \in \tilde{y}$ such that $\tilde{y} \in \hat{W}$:

$$\mathcal{M}_{F_c(\Gamma)}, x \models Cp, \text{ if } p \text{ is clearly visible at } \hat{w}.$$

To construct function $N_{F_c(\Gamma)}$ that shows the concept of focus as modality, we take the following two steps:

1. Constructing function $N_{F_c(\Gamma)}^{\hat{w}}$ for each $\hat{w} \in \hat{W}$.
2. Combining all $N_{F_c(\Gamma)}^{\hat{w}}$.

First, we define function $N_{F_c(\Gamma)}^{\hat{w}}$.

Definition 11.4 For each $\hat{w} \in \hat{W}$, a function $N_{F_c(\Gamma)}^{\hat{w}} : \bigcup \hat{w} \to 2^{2^w}$ is defined by:

$$N_{F_c(\Gamma)}^{\hat{w}} = \{\bigcup A \mid A \subseteq U(\hat{w})\}, \forall x \in \bigcup \hat{w},$$

where $U(\hat{w}) = (\tilde{W} \backslash \hat{w}) \cup (\bigcup \hat{w})$. If $A = \emptyset$, then $\bigcup A = \emptyset$.

Combining all functions $N_{F_c(\Gamma)}^{\hat{w}}$, we define function $N_{F_c(\Gamma)}$.

Definition 11.5 For all $x \in W$, function $N_{F_c(\Gamma)}^{\hat{w}} : W \to 2^{2^w}$ is defined by:

$$N_{F_c(\Gamma)}(x) = \begin{cases} N_{F_c(\Gamma)}^{\hat{w}}(x), & \text{if } F_c(\Gamma) \neq \emptyset \text{ and } x \in \bigcup \hat{w}, \\ \{W, \emptyset\}, & \text{otherwise.} \end{cases}$$

Function $N_{F_c(\Gamma)}$ is easily confirmed to be well-defined by Definition 11.5. The key to this construction is set $U(\hat{w})$, which provides "unit" of construction at each possible world $x \in \bigcup \hat{w}$.

$U(\hat{x})$ does not contain any equivalent classes of possible worlds $\tilde{y}$ in $\hat{w}$ because we must capture the property that atomic sentence p is visible iff p is true at *all* equivalent classes of possible worlds in $\hat{w}$ or false at *all* equivalent classes of possible worlds in $\hat{w}$. If some $\tilde{y} \in \hat{w}$ are contained in $U(\hat{w})$, atomic sentence $q \in V_s(\Gamma) \backslash F_c(\Gamma)$ may become "clearly visible". Therefore, no $\tilde{y} \in \hat{w}$ should be included in $U(\hat{w})$.

$N_{V_s(\Gamma)}$ and $N_{F_c(\Gamma)}$ differs as follows: (1) $N_{V_s(\Gamma)}$ treats all combinations of unions of equivalent classes in $\tilde{W}$ as a "unit" of construction, while $N_{F_c(\Gamma)}$ treat some restricted parts of combinations of equivalent classes in $\tilde{W}$ because we must distinguish between clearly visible sentences and obscurely visible sentences using function $N_{F_c(\Gamma)}$, and concept p is *clearly visible*-or is *in focus*-that p is either $\{\mathbf{T}\}$ or $\{\mathbf{F}\}$ at *all* equivalent classes in $\hat{w}$. (2) $N_{F_c(\Gamma)}$ must treat case $F_c(\Gamma) = \emptyset$ corresponding to the situation in which *nothing is clear*.

$N_{F_c(\Gamma)}$ satisfies the following conditions:

Lemma 11.3 *Constructed function $N_{F_c(\Gamma)}$ satisfies conditions (c), (n), and (v!).*

We omit the proof of Lemma 11.3 because it is essentially the same sa the proof of Lemma 11.1.

In general, however, function $N_{F_c(\Gamma)}$ does not satisfy conditions (4) and (5). As with case $N_{V_s(\Gamma)}$, $N_{F_c(\Gamma)}$ also does not satisfy (m), (d) and (t).

Using $N_{F_c(\Gamma)}$, we construct Scott-Montague model $M_{F_c(\Gamma)} = \langle W, N_{F_c(\Gamma)}, v \rangle$. Lemma 11.3 indicates that model $M_{F_c(\Gamma)}$ validates schemata **C**, **N**, and **V!**, and all properties shown in Lemma 11.2 are also valid for operator C, i.e., if both p and q are clearly visible, then $\neg p$, $p \wedge q$, $p \vee q$ and $p \rightarrow q$ are also clearly visible.

Lemma 11.4 *1. If $p \in V_s(\Gamma)$, then $M_{F_c(\Gamma)}, x \models Cp$ for all $w \in W$.*
2. For any $p \in \mathcal{L}(\mathcal{P})$ and all $w \in W$, if $M_{F_c(\Gamma)}, x \models Cp$, then $M_{F_c(\Gamma)}, x \models C\neg p$.
3. For any $p, q \in\in \mathcal{L}(\mathcal{P})$ and all $w \in W$, if both $M_{F_c(\Gamma)}, x \models Cp$ and $M_{F_c(\Gamma)}, x \models Cq$, then $M_{F_c(\Gamma)}, x \models C(p \wedge q)$, $M_{F_c(\Gamma)}, x \models C(p \vee q)$, and $M_{F_c(\Gamma)}, x \models C(p \rightarrow q)$.

As with Lemma 11.3, we omit the proof of Lemma 11.4. For Lemmas 11.3 and 11.4, we have the following result:

Theorem 11.2 *Let Γ be a non-empty set of non-modal sentences. $\hat{W}$ be the quotient set of set $\tilde{W}$ based on $F_c(\Gamma)$, and $M_{F_c(\Gamma)} = \langle W, N_{F_c(\Gamma)}, v \rangle$ be a Scott-Montague model that has function $N_{F_c(\Gamma)}$ by Definition 11.5. For any non-modal sentence p if p is clearly visible at $\hat{w} \in \hat{W}$, then $M_{F_c(\Gamma)}, x \models Cp$ for all $x \in \tilde{y}$ such that $\tilde{y} \in \hat{w}$.*

As with Vp, the converse of Theorem 11.2 is not always satisfied.

Example 11.3 We use the same setting of Example 11.1. Using $\hat{w}_1 = \{\tilde{w}_1, \tilde{w}_5\} = \{\{w_1, w_2\}, \{w_5, w_6\}\}$, $\hat{w}_3 = \{\tilde{w}_3, \tilde{w}_7\} = \{\{w_3, w_4\}, \{w_7, w_8\}\}$, we get sets $U(\hat{w}_1)$ as follows:

$$U(\hat{w}_1) = \{\{w_1, w_2.w_5, w_6\}, \{w_3, w_4\}, \{w_7, w_8\}\},$$
$$U(\hat{w}_3) = \{\{w_3, w_4.w_7, w_8\}, \{w_1, w_2\}, \{w_5, w_6\}\}.$$

Using $N_{F_c(\Gamma)}^{\hat{w}_1}$ and $U(\hat{w}_3)$, we construct functions $N_{F_c(\Gamma)}^{\hat{w}_1}$ and $N_{F_c(\Gamma)}^{\hat{w}_3}$ by Definition 11.4, and Scott-Montague model $M_{F_c(\Gamma)} = \langle W, N_{F_c(\Gamma)}, v \rangle$ with function $N_{F_c(\Gamma)}$ by Definition 11.5.

We have $V_s(\Gamma) = \{p, q\}$ and $F_c(\Gamma) = \{q\}$, thus q is clearly visible but p is obscurely visible. Here, for model $M_{F_c(\Gamma)}$, it holds that $\|q\|^{M_{F_c(\Gamma)}} = \{w_1, w_2, w_5, w_6\} \in N_{F_c(\Gamma)}$ for all $x \in W$. For example, $M_{F_c(\Gamma)}, w_2 \models Cq$ and $M_{F_c(\Gamma)}, w_2 \models C\neg q$. We cannot construct $\|p\|^{M_{F_c(\Gamma)}}$ at either $\hat{w}_1$ or $\hat{w}_3$, so we have $M_{F_c(\Gamma)}, w_i \not\models Cp$ and $M_{F_c(\Gamma)}, w_i \not\models C\neg p$.

We also treat complex non-modal sentences. In Example 11.1, $p \rightarrow q$ is clearly visible at $\hat{w}_1$, but obscurely visible at $\hat{w}_3$. The following holds for all $w_i \in \bigcup \hat{w}_i$:

$$\|p \rightarrow q\|^{M_{F_c(\Gamma)}} = \{w_1, w_2, w_5, w_6, w_7.w_8\}$$
$$= \{w_1, w_2, w_5, w_6\} \cup \{w_7, w_8\}$$
$$\in N_{F_c(\Gamma)}(w_i).$$

We thus have, for example, $M_{F_c(\Gamma)}, w_6 \models C(p \rightarrow q)$. We cannot, however, construct $\|p \rightarrow q\|^{M_{F_c(\Gamma)}}$ from elements in $U(\hat{w}_3)$, so $\|p \rightarrow q\|^{M_{F_c(\Gamma)}} \notin N_{M_{F_c(\Gamma)}}(w_j)$ for any $w_j \in \bigcup \hat{w}_3$, and therefore, for example, $M_{F_c(\Gamma)}, w_3 \not\models C(p \rightarrow q)$ holds.

11.6 Conclusions

We represented the concepts of visibility and focus as modalities. Modal operator V we proposed, meaning visible, involves function $N_{V_s(\Gamma)}$ to capture properties of visibility. Scott-Montague models $\mathcal{M}_{V_s(\Gamma)}$ we propose demonstrates that if p is visible at $\tilde{w}$, then $\mathcal{M}_{V_s(\Gamma)}, x \models Vp$ for all $x \in \tilde{w}$.

Another modal operator C we proposed, meaning clearly visible, involves constructing function $N_{F_c(\Gamma)}$. Another Scott-Montague model $\mathcal{M}_{V_c(\Gamma)}$ that we proposed demonstrates that if p is clearly visible, at $\tilde{w}$, then $\mathcal{M}_{V_c(\Gamma)}, x \models Cp$ for all $x \in \tilde{y}$ such that $\tilde{y} \in \hat{w}$.

Among projected issues that remain, we must first explore connections between V and C via multi modal Scott-Montague models, and axiomatically characterize visibility and focus. Combining with other modal logic, especially, logic of knowledge and belief [2], and logic of time [14] are also of interest. We must also consider relationships among our framework and zooming reasoning systems [9–11].

References

1. Chellas, B.: Modal Logic: An Introduction. Cambridge University Press, Cambridge (1980)
2. Hintikka, S.: Knowledge and Belief. Cornell University Press, Ithaca (1962)
3. Kleene, S.: Introduction to Metamathematics. North-Nostrand, Amsterdam (1952)
4. Kudo, Y., Murai, T.: Visibility and focus: an extende framework for granular reasoning. In: Nakamatsu, K., Abe, J.M. (eds.) Advances in Logic Based Intelligent Systems, pp. 280–287. IOS Press (2005)
5. Kudo, Y., Murai, T.: A note on granular reasoning and semantics of four-valued logics. AIP Conf. Proc. **819**, 453–460 (2006)
6. Kudo, Y., Murai, T.: A modal characterization of granular reasoning based on Scott-Montague models. In: Proceedings of SCIS& ISIS2008, pp. 991–995 (2008)
7. Lin, T.Y.: Granular computing on binary relation, I data mining and neighborhood systems, II rough set representations and belief functions. In: Polkowski, L., Skowron, A. (eds.) Rough Sets in Knowledge Discovery 1: Methodology and Applications, pp. 107–121, 122–140. Physica-Verlag, Heidelberg (1998)
8. Murai, T., Nakata, M., Sato, Y.: A note on filtration and granular reasoning. In: Terano, T. et al. (eds.) New Frontiers in Artificial Intelligence, LNAI 2253, pp. 385–389. Springer (2001)
9. Murai, T., Resconi, G., Nakata, M. and Sato, Y.: Operations of zooming In and out on possible worlds for semantic fields. In: Damiani, E. et al. (eds.) Knowledge-Based Intelligent Information Engineering Systems and Allied Technologies, pp. 1083–1087 (2002)
10. Murai, T., Resconi, G., Nakata, M., Sato, Y.: Granular reasoning using zooming in & out: Part 1. Propositional Reasoning. In: Proceedings of International Conference on Rough Sets, Fuzzy Sets, Data Mining, and Granular Computing (2003)
11. Murai, T., Sato, Y., Resconi, G., Nakata, M.: Granular reasoning using zooming in & out: Part2. Aristotle's categorial syllogism. Electron. Notes Theor. Comput. Sci. **82**, 186–197 (2003)
12. Pawlak, Z.: Rough sets. Sci. Int. J. Comput. Inf. **11**, 341–356 (1982)
13. Pawlak, Z.: Rough Sets: Theoretical Aspects of Reasoning about Data. Kluwer, Dordrecht (1991)
14. Prior, A.: Past, Present and Future. Clarendon Press, Oxford (1967)
15. Skowron, A.: Toward intelligent systems: calculi of information granules. In: Terano, T. et al. (eds.) New Frontiers in Artificial Intellifgence, LNAI 2253, pp. 251–260. Springer (2001)

Chapter 12
Directions for Future Work in Rough Set Theory

Abstract This chapter concludes this book by sketching some directions for future work. In the context of soft computing dealing with vagueness, important problems in rough set theory include decision tables, decision rules, attribute reduction, etc. While some of these issues have been discussed in this book, we consider them further. There are several generalizations of rough set theory and unifications with other theories. Thus, we discuss several issues about generalizations and unifications. A theory of three-way decisions is also presented as a promising framework for generalizing rough set theory.

Keywords Rough set theory · Decision table · Decision-theoretic rough set · Three-way decisions

12.1 Furthes Issues in Decision Tables and Decision Rule

This chapter concludes this book by sketching some directions for future work. In the context of soft computing dealing with vagueness, important problems in rough set theory include decision tables, decision rules, attribute reduction, etc. While some of these issues have been discussed in this book, we consider them further.

There are several generalizations of rough set theory and unifications with other theories, and we also discuss several issues about generalizations and unifications. A theory of three-way decisions is also presented as a promising framework for generalizing rough set theory.

We have already addressed further issues concerning logical foundations for rough set theory in Akama, Murai and Kudo [1]. Part of the discussions below may overlap those given before in our previous book. In the present section, we deal with several issues in other aspects of rough set theory.

The concept of *rough set* was originally introduced by Pawlak [13, 14] to model vagueness. It is well known that *fuzzy set theory* was also developed by Zadeh [19] to model vagueness. However, the concepts of vagueness in these theories should be distinguished. Fuzzy set theory deals with gradualness of knowledge by using fuzzy

© Springer Nature Switzerland AG 2020
S. Akama et al., *Topics in Rough Set Theory*, Intelligent Systems
Reference Library 168, https://doi.org/10.1007/978-3-030-29566-0_12

membership, whereas rough set theory deals with granularity of knowledge by using indiscernibility relation under background knowledge.

This means that rough set theory and fuzzy set theory are different approaches to vagueness, but they are complementary. It would be thus natural to study further issues from various (extended to unified) perspectives.

In this section, some problems of decision tables and decision rules are discussed. After rough set theory was developed in 1982, we can find a rich literature on applications of rough set theory, in particular, data analysis. The reason would be that there is some connection between set-theoretic approximation of concepts, in particular, extraction of lower approximations, and extraction of if-then type rules.

Let U be a finite set of objects in question and R be an *equivalence relation* on U for representing background knowledge. For $X \subseteq U$, we define *lower approximation* $\underline{R}(X)$ and *upper approximation* $\overline{R}(X)$ as follows:

$$\underline{R}(X) = \{x \in U \mid [x]_R \subseteq X\},$$
$$\overline{R}(X) = \{x \in U \mid [x]_R \cap X \neq \emptyset\}.$$

Here, set $[x]_R$ is an equivalence class of x by indiscernibility relation R. A pair $(\underline{R}(X), \overline{R}(X))$ is called the *rough set* with respect to X.

A table form of data for analysis is called a *decision table*. Attributes (items) in a decision table are classified as *decision attributes* which correspond to attributes we want to know their properties and *condition attributes* which correspond to attributes describing their properties.

A decision table is defined as $(U, C \cup \{d\}, V, \rho)$, where U is finite set of objects, C is a set of condition attributes, $d \notin C$ is a decision attribute, V is a set of values of attributes, and $\rho : U \times (C \cup \{d\}) \to V$ is a function which assigns the value $\rho(x, a) \in V$ of each attribute $a \in C \cup \{d\}$ to each object $x \in U$.

$$R_A = \{(x, y) \in U \times U \mid \rho(x, a) = \rho(y, a), \forall a \in A\} \qquad (12.1)$$

Indiscernibility relation is equivalence relation (cf Pawlak [14]). An equivalence class $D_1, \ldots, D_n$ obtained by decision attributes with indiscernibility relation is called a *decision class* which is used as the concept under approximations.

Consider indiscernibility relation R_A based on a subset of condition attributes $A \subseteq U$. Since $\underline{R}_A(D_j) \subseteq D_j$ for any decision class D_j ($1 \leq j \leq n$), every element of lower approximation $\underline{R}_A(D_j)$ (unless it is empty set) belong to decision class D_j.

For any object $a \in \underline{R}_A(D_j)$, $A = \{a_1, \ldots, a_k\} \subseteq C$, decision attribute d, inclusion relation $\underline{R}_A(D_j) \subseteq D_j$ means the rule saying that if the value of a_1 is $\rho(x, a_1)$ and ... and the value of ask is $\rho(x, a_k)$ then the value of d is $\rho(x, d)$. The rule of the form is called a *decision rule*, expressed as:

$$\bigvee_{d_i \in A} [a_i = \rho(x, a_1)] \to [d = \rho(x, d)] \qquad (12.2)$$

Here, the left side of $\to$ is called the *antecedent* and the right side the *consequent*, respectively.

The decision rule obtained from the objects included in lower approximation of decision class is a consistent rule in which the consequent is uniquely decided for the antecedent. It is also possible to create the decision rule obtained from the objects not included in lower approximation of decision class, and such a decision rule is an inconsistent rule in that the consequent is not unique for the antecedent.

As the description of regularity in a decision table, shorter antecedent of a decision rule is desirable as a rule. On the other hand, classification capability of objects by decision rule must have the same capability when using all condition attributes. Computing subset of condition attributes giving the antecedent of a decision rule is called *attribute reduction* and the resulting subset of condition attributes is called *relative reduct*, respectively.

One of the methods for extracting all relative reducts from a decision table is to use the so-called a *discernibility matrix*; see Skwron and Rauszer [15]. It is proved, however, in Skwron and Rauszer [15] that the problem of extracting all relative reducts from a given decision table is NP-hard.

The fact implies that it is difficult to obtain all relative reducts in large-scale databases. Thus, some alternative methods have been worked out. There are several such approaches in the literature. For example, Chouchoulas and Shen proposed an algorithm of finding heuristically several candidates of relative reducts, i.e., *Quacked* (cf. [2]); also see Hu et al. [6, 7], Kudo and Murai [9].

Another approach is to extract many relative reducts from data with many values of attributes as in Kudo and Murai [10]. Observe that, as shown in Chap. 3, the idea of discernibility matrix can be also applied to object reduction.

12.2 Generalizations and Unifications

Many concepts used in rough set theory can be generalized or unified with other related approaches. It is in fact natural to explore extensions and variants of rough set theory, and unifications with other theories. Here, we shortly discuss some of them.

An interesting generalization is *variable precision rough set* (VPRS) due to Ziarko [20]. VPRS model is a generalization of Pawlak rough set model to model classification problems with uncertain or imprecise information. In other words, VPRS model allows for some degree of misclassification in the largely correct classification.

Although standard rough set models are based on equivalence relation, we can use any binary relation. One idea is to formalize generalized rough set models by interpreting lower and upper approximations by Kripke semantics for modal logic; see Yao and Lin [17]. Another idea was developed by Pagliani [12] who constructs rough set models by means of Nelson algebras.

As a set of equivalence classes becomes a classification of a given set, it is natural to work out generalized classifications. Examples include *neighborhood rough set* due to Lin [11] and *covering-based rough set* due to Zhu [21]. Note that such generalizations are compatible with granular computing.

In standard rough set theory, indiscernibility relation is used to compare objects described by certain attributes. We can relax indiscernibility relation by using other relations. Grecco et al. [5] proposed the so-called *dominance-based rough set* based on *dominance relation*. Dominance-based rough set can model the preference orders of attribute domains and semantic correlation of attributes.

Based on Pawlak's rough set theory, in a decision table null values are not allowed. However, it is possible to develop a rough set theory for decision tables with null values. For example, Kryzskiewica [8] invesigates incomplete information systems.

It is an interesting topic to unifying rough set theory and fuzzy set theory, since vagueness considered in these theories are different. Thus, unifying these two theories seem to yield more powerful theory for vagueness. For example, Dubois and Prade [3] proposed *rough fuzzy sets* and *fuzzy rough sets*.

Modal logic can serve as a logical foundation for rough set theory, since necessity and possibility operators in modal logic are closely related to approximations in rough set theory. Yao and Lin [17] systematically studied a modal logic approach to rough set theory, which generalize Pawlak rough set theory. In the modal logic approach, other semantics like neighborhood semantics and Kripke semantics with non-normal worlds would further generalize rough set theory.

12.3   Three-Way Decisions

Finally, we shortly present the outline of a theory of *three-way decisions*, which can generalize several set theories including rough set theory. A theory of three-way decisions, as proposed by Yao [16], is constructed based on the notions of acceptance, rejection and noncommitment. It is an extension of the commonly used binary-decision model with an added third option.

The problem of three-way decisions is described as follows. Suppose U is a finite non-empty set and C is a finite set of criteria. The problem of three-way decisions is to divide, based on the set of criteria C, U into three pair-wise disjoint regions, POS, NEG, and BND, called the positive, negative, and boundary regions, respectively.

Three-way decisions play a key role in everyday decision-making and have been widely used in many fields and disciplines. The concept of three-way decisions provides an appealing interpretation of three regions in probabilistic rough sets. The positive and negative regions are sets of accepted objects and rejected objects, respectively. The boundary region is the set of objects for which neither acceptance nor rejection is possible, due to uncertain or incomplete information.

A close examination of studies and applications of three-way decisions shows that:

(a) Essential ideas of three-way decisions are general applicable to a wide range of decision-making problems.

(b) We routinely make three-way decisions in everyday life.

(c) Three-way decisions appear across many fields and disciplines.

(d) There is a lack of formal theory for three-way decisions.

These findings motivate a study of a theory of three-way decisions in its own right. We expect to see further developments in connection with many applications.

We now elaborate on three-way decisions. Three-way decisions improve traditional two-way ones. The basic idea of three-way decisions is to use a ternary classification according to evaluation of a set of criteria. There are in fact two difficulties with two-way decisions. One is their stringent binary assumption of satisfiability of objects. The other is the requirement of a dichotomous classification.

Three-way decisions use thresholds on the degrees of satisfiability to make one three decisions:

(a) accept an object as satisfying the set of criteria if its degree of satisfiability is at or above a certain level;

(b) reject the object by treating it as not satisfying the criteria if its degree of satisfiability is at or below another level;

(c) neither accept nor reject the object but opt for a noncommitment.

Therefore, the problem of three-way decisions can be formally described as follows: Suppose U is a finite non-empty set and C is a finite set of criteria. The problem of three-way decisions is to divide, based on the set of criteria C, U into three pair-wise disjoint regions POS, NEG and BND, called the positive, negative, and boundary regions, respectively.

Corresponding to the three regions, one may construct rules for three-way decisions. Yao showed a theory of three-way decisions can describe and explain three-way decisions in many-valued logics and decision-theoretic rough sets.

For the description of the satisfiability of objects, rules for acceptance and rules for rejection, the notion of evaluation of objects and designated values for acceptance and designated values for rejections are needed. Evaluations are needed for the degrees of satisfiability, and designated values for acceptance are acceptable degrees of satisfiability, and designated values for rejection are acceptable degrees of non-satisfiability. Based on these concepts, a theory of three-way decisions can be formalized.

Evaluations for acceptance and rejection can be constructed based on the set of criteria. Thus, we can present evaluation-based three-way decisions. Here, the formalization of Pawlak rough sets are shown. For showing it, we need a pair of evaluations, i.e., one for the purpose of acceptance and the other for rejection.

Definition 12.1 Let U be a finite non-empty set and $(L_a, \preceq_a)$, $(L_r, \preceq_r)$ are two posets (partially ordered sets). A pair of functions $v_a : U \rightarrow L_a$ and $v_r ; U \rightarrow L_r$ is called an acceptance evaluation and a rejection evaluation, respectively. For $x \in U$, $v_a(x)$ and $v_r(x)$ are called acceptance and rejection values of x, respectively.

To accept an object, its value $v_a(x)$ must be in a certain subset of L_a representing the acceptance region of L_a, and to reject an object, its value $v_r(x)$ must be in a subset if L_r representing the rejection region of L_r, respectively. These values are called *designated values*.

Based on the two sets of designated values, we can obtain three regions for three-way decisions.

Definition 12.2 Let $\emptyset \neq L_a^+ \subseteq L_a$ be a subset of L_a called the designated values for acceptance, and $\emptyset \neq L_r^- \subseteq L_r$ be a subset of L_r called the designated values for rejection. The positive, negative and boundary regions of three-way decisions induced by (v_a, v_r) are defined by

$$POS_{L_a^+, L_r^-}(v_a, v_r) = \{x \in U \mid v_a(x) \in L_a^+ \text{ and } v_r(A) \notin L_r^-\},$$
$$NEG_{L_a^+, L_r^-}(v_a, v_r) = \{x \in U \mid v_a(x) \notin L_a^+ \text{ and } v_r(A) \in L_r^-\},$$
$$BND_{L_a^+, L_r^-}(v_a, v_r) = (POS_{L_a^+, L_r^-}(v_a, v_r) \cup NEG_{L_a^+, L_r^-}(v_a, v_r))^c.$$

Recall that Pawlak rough set theory deals with approximation of a concept based on a family of definable concepts. Let $E \subseteq U \times U$ denote an equivalence relation on U. The equivalence class containing x is defined by $[x]_E = [x] = \{y \in Y \mid xRy\}$.

For a subset $A \subseteq U$, the Pawlak rough set lower and upper approximation of A are defined.

$$\underline{apr}(A) = \{x \in U \mid [x] \subseteq A\},$$
$$\overline{apr}(A) = \{x \in U \mid [x] \cap A \neq \emptyset\}.$$

Then, we define the Pawlak positive, negative and boundary regions as follows:

$$POS(A) = \underline{apr}(A),$$
$$NEG(A) = U - \overline{apr}(A),$$
$$BND(A) = \overline{apr}(A) - \underline{apr}(A).$$

These regions are defined by set inclusion, and they are pairwise disjoint. Conversely, we can compute the pair of approximations by:

$$\underline{apr}(A) = POS(A),$$
$$\overline{apr}(A) = POS(A) \cup BND(A).$$

Therefore, rough set theory can be formalized by either a pair of approximations or three regions.

Three-way decisions with rough sets can be formulated as follows. Let $L_a = L_r = \{F, T\}$ with $F \leq T$ and let $L_a^+ = L_r^- = \{T\}$. All objects in the same equivalence class have the same description.

Based on descriptions of objects, we have a pair of an acceptance evaluation and a rejection evaluation as follows:

$$v_{(a,A)}(x) = \begin{cases} T & ([x] \subseteq A), \\ F & (\neg([x] \subseteq A)) \end{cases} \quad v_{(r,A)}(x) = \begin{cases} T & ([x] \subseteq A^c), \\ F & (\neg([x] \subseteq A^c)) \end{cases} \quad (12.3)$$

By Definition 11.2, for a set $A \subseteq U$, there are the following three-way decisions.

$$POS_{(\{T\},\{T\})}(A) = \{x \in U \mid v_{(a,A)}(x) \in \{T\} \text{ and } v_{(r,A)}(x) \notin \{T\}\}$$
$$= \{x \in U \mid v_{(a,A)}(x) = T\}$$
$$= \{x \in U \mid [x] \subseteq A\}$$
$$NEG_{(\{T\},\{T\})}(A) = \{x \in U \mid v_{(a,A)}(x) \notin \{T\} \text{ and } v_{(r,A)}(x) \in \{T\}\}$$
$$= \{x \in U \mid v_{(r,A)}(x) = T\}$$
$$= \{x \in U \mid [x] \subseteq A^c\}$$
$$BND_{(\{T\},\{T\})}(A) = (POS_{(v_a,v_r)}(A) \cup NEG_{(v_a,v_r)}(A))^c$$
$$= \{x \in U \mid \neg([x] \subseteq A) \text{ and } \neg([x] \subseteq A^c)\}$$

The above is a reformulation of Pawlak rough set in three regions based uniformly on set inclusion. It provides additional insights into rough set approximation. It explicitly shows that acceptance is based on an evaluation of the condition $[x] \subseteq A$ and rejection is based on an evaluation of the condition $[x] \subseteq A^c$.

By the two conditions, both decisions of acceptance and rejection are made without any error. Whenever there is any doubt, namely, $\neg([x] \subseteq A)$ and $\neg([x] \subseteq A^c)$, a decision of noncommitment is made.

For a total order, it is possible to define the sets of designated values for acceptance and rejection by a pair of thresholds.

Definition 12.3 Suppose $(L, \preceq)$ is a totally ordered set, that is, $\preceq$ is a total order. For two elements $\alpha, beta$ with $\beta \prec \alpha$ (i.e., $\beta \preceq \alpha$ and $\neg(\alpha \preceq \beta)$), suppose that the set of designated values for acceptance is given by $L^+ = \{t \in\mid t \succeq \alpha\}$ and the set of designated values for rejection is given by $L^- = \{b \in L \mid b \preceq \beta\}$. For an evaluation function $v : U \to L$, its three regions are defined by:

$$POS_{(\alpha,\beta)}(v) = \{x \in U \mid v(x) \succeq \alpha\}$$
$$NEG_{(\alpha,\beta)}(v) = \{x \in U \mid v(x) \preceq \beta\}$$
$$BND_{(\alpha,\beta)}(v) = \{x \in U \mid \beta \prec v(x) \prec \alpha\}$$

Recall that *decision-theoretic rough sets* (DTRS) of Yao et al. [18] are a quantitative generalization of Pawlak rough sets. As discussed Chap. 2, it considers the degree of inclusion of an equivalence class in a set. The acceptance-rejection evaluation used by a DTRS model is the conditional probability $v_A(x) = Pr(A \mid [x])$, with values from the totally ordered set of $([0, 1], \leq)$.

In DTRS model, the required parameters for defining probabilistic lower and upper approximations are calculated based on more familiar notions of costs (risks) through the well-known *Bayesian decision procedure*. The Bayesian decision procedure (cf. [4]) deals with making decision with minimum risk based on observed evidence.

Let $\Omega = \{w_1, \ldots, w_s\}$ be a finite set of s states, and let $\mathcal{A} \{a_1, \ldots, a_m\}$ be a finite set of m possible actions. Let $P(w_j \mid \mathbf{x})$ be the conditional probability of an object x being in state w_j given that the object is described by $\mathbf{x}$. Let $\lambda(a_i \mid m_j)$ denote the loss, or cost, for taking action a_i when the state is w_j. For an object with description $\mathbf{x}$, suppose action a_i is taken.

Since $P(w_j \mid \mathbf{x})$ is the probability that the true state is w_j given $\mathbf{x}$, the expected loss associated with taking action a_i is given by:

$$R(a_i \mid \mathbf{x}) = \sum_{j=1}^{s} \lambda(a_i \mid w_j) P(w_j \mid \mathbf{x})$$

The quantity $R(s_i \mid \mathbf{x})$ is also called the *conditional risk*.

Given a description $\mathbf{x}$, a decision rule is a function $\tau(\mathbf{x})$ that specifies which action to take. That is, for every $\mathbf{x}$, $\tau(\mathbf{x})$ takes one of the actions, $a_1, \ldots, a_m$. The overall risk $\mathbf{R}$ is the expected loss associated with action $\tau(\mathbf{x})$, the overall risk is defined by:

$$\mathbf{R} = \sum_{\mathbf{x}} R(\tau(\mathbf{x}) \mid \mathbf{x}) P(\mathbf{x})$$

where the summation is over the set of all possible descriptions of objects. If $\tau(\mathbf{x})$ is chosen so that $R(\tau(\mathbf{x} \mid \mathbf{x})$ is as small as possible for every $\mathbf{x}$, the overall risk $\mathbf{R}$ is minimized.

In an approximation space $apr = (U, E)$, an equivalence class $[x]$ is considered to be the description of x. The partition U/E is the set of all possible descriptions. The classification of objects according to approximation operators can be easily fitted into the Bayesian decision-theoretic framework. The set of states is given by $\Omega = \{A, A^c\}$ indicating that an element is in A and not in A, respectively.

We use the same symbol to denote both a subset A and the corresponding state. With respect to the three regions, the set of actions is given by $\mathcal{A} = \{a_1, a_2, a_3\}$ where a_1, a_2 and a_3 represent the three actions in classifying an object, deciding $POS(A)$, deciding $NEG(A)$, and deciding $BND(A)$, respectively.

Let $\lambda(a_i \mid A)$ denote the loss incurred for taking action a_i when an object belong to A, and let $\lambda(a_i \mid A^c)$ denote the loss incurred for taking the same action when the object does not belong to A.

$$R(a_1 \mid [x]) = \lambda_{11} P(A \mid [x]) + \lambda_{12} P(A^x \mid [x]),$$
$$R(a_2 \mid [x]) = \lambda_{21} P(A \mid [x]) + \lambda_{22} P(A^x \mid [x]),$$
$$R(a_3 \mid [x]) = \lambda_{31} P(A \mid [x]) + \lambda_{32} P(A^x \mid [x]),$$

where $\lambda_{i1} = \lambda(a_i \mid A)$, $\lambda_{i2} = \lambda(a_i \mid A^c)$ $(i = 1, 2, 3)$. The Bayesian decision procedure leads to the following minimum-risk decision rules:

(P) If $R(a_1 \mid [x]) \leq R(a_2 \mid [x])$ and $R(a_1 \mid [x]) \leq R(a_3 \mid [x])$, decide $POS(A)$,
(N) If $R(a_2 \mid [x]) \leq R(a_1 \mid [x])$ and $R(a_2 \mid [x]) \leq R(a_3 \mid [x])$, decide $NEG(A)$,
(B) If $R(a_3 \mid [x]) \leq R(a_1 \mid [x])$ and $R(a_3 \mid [x]) \leq R(a_2 \mid [x])$, decide $BND(A)$.

Tie-breaking criteria should be added so that each element is classified into only one region. Since $P(A \mid [x]) + P(A^c \mid [x]) = 1$, we can simplify the rules to classify any object in $[x]$ based only on probabilities $P(A \mid [x])$ and the loss function λ_{ij} ($i = 1, 2, 3, j = 1, 2$).

Lin gave several examples of decision rules according to the conditions for a loss function. For example, consider a special kind of loss function with $\lambda_{11} \leq \lambda_{31} < \lambda_{21}$ and $\lambda_{22} \leq \lambda_{32} < \lambda_{12}$. That is, the loss of classifying an object x belonging to A into the positive region $POS(A)$ is less than or equal to the loss of classifying x into the boundary region $BND(A)$, and both of those losses are strictly less than the loss of classifying x into the negative region $NEG(A)$.

The reverse order of losses is used for classifying an object that does not belong to A. For this type of loss function, the minimum-risk decision rules (P)–(B) can be written as;

(P) If $P(A \mid [x]) \geq \gamma$ and $P(A \mid [x]) \geq \alpha$, decide $POS(A)$,
(N) If $P(A \mid [x]) \leq \beta$ and $P(A \mid [x]) \leq \gamma$, decide $NEG(A)$
(B) If $\beta \leq P(A \mid [x]) \leq \alpha$, decide $BND(A)$

where

$$\alpha = \frac{\lambda_{12} - \lambda_{32}}{(\lambda_{31} - \lambda_{32}) - (\lambda_{11} - \lambda_{12})}$$
$$\gamma = \frac{\lambda_{12} - \lambda_{22}}{(\lambda_{21} - \lambda_{22}) - (\lambda_{11} - \lambda_{12})}$$
$$\beta = \frac{\lambda_{32} - \lambda_{22}}{(\lambda_{21} - \lambda_{22}) - (\lambda_{31} - \lambda_{32})}$$

By the assumptions, $\lambda_{11} \leq \lambda_{31} < \lambda_{21}$ and $\lambda_{22} \leq \lambda_{32} < \lambda_{12}$, it follows that $\alpha \in (0, 1]$, $\gamma \in (0, 1)$, and $\beta \in [0, 1)$.

For a loss function with $\lambda_{11} \leq \lambda_{31} < \lambda_{21}$ and $\lambda_{22} \leq \lambda_{32} < \lambda_{12}$, more results about the required parameters α, β and γ are summarized as follows:

1. If a loss function satisfies the condition $(\lambda_{12} - \lambda_{32})(\lambda_{21} - \lambda_{31}) \geq (\lambda_{31} - \lambda_{11})(\lambda_{32} - \lambda_{22})$ then $\alpha \geq \gamma \geq \beta$.
2. If a loss function satisfies the condition $(\lambda_{12} - \lambda_{22})(\lambda_{31} - \lambda_{11})$, then $\alpha \geq 0.5$.
3. If a loss function satisfies the conditions $(\lambda_{12} - \lambda_{22})(\lambda_{31} - \lambda_{11})$ and $(\lambda_{12} - \lambda_{32})(\lambda_{21} - \lambda_{31}) \geq (\lambda_{31} - \lambda_{11})(\lambda_{32} - \lambda_{22})$ then $\alpha \geq 0.5$ and $\alpha \geq \beta$.
4. If a loss function satisfies the condition $(\lambda_{12} - \lambda_{32})(\lambda_{32} - \lambda_{22}) = (\lambda_{31} - \lambda_{11})(\lambda_{31} - \lambda_{11})(\lambda_{21} - \lambda_{31})$, then $\beta = 1 - \alpha$.
5. If a loss function satisfies the two sets of equivalent conditions,
   (i) $(\lambda_{12} - \lambda_{32})(\lambda_{21} - \lambda_{31}) \geq (\lambda_{31} - \lambda_{11})(\lambda_{32} - \lambda_{22})$,
       $(\lambda_{12} - \lambda_{32})(\lambda_{32} - \lambda_{22}) = (\lambda_{31} - \lambda_{11})(\lambda_{21} - \lambda_{31})$,
   (ii) $\lambda_{12} - \lambda_{32} \geq \lambda_{31} - \lambda_{11}$, $(\lambda_{12} - \lambda_{32})(\lambda_{32} - \lambda_{22}) = (\lambda_{31} - \lambda_{11})(\lambda_{21} - \lambda_{31})$,
   then $\alpha = 1 - \beta \geq 0.5$.

Here, the condition of Case 1 guarantees that the probabilistic lower approximation of a set is a subset of its probabilistic upper approximation. The condition of Case 2 ensures that a lower approximation of A consists of those elements whose majority equivalent elements are in A. The condition of Case 4 results in a pair of dual lower and upper approximation operators. Case 3 is a combination of Cases 1 and 2. Case 5 is the combination of Cases 1 and 4 or the combination of Cases 3 and 4.

When $\alpha > \beta$, we have $\alpha > \gamma > \beta$. After tie-breaking, we obtain the decision rules:

(P1) If $P(A \mid [x]) \geq \alpha$, decide $POS(A)$,
(N1) If $P(A \mid [x]) \leq \beta$ decide $NEG(A)$
(B1) If $\beta < P(A \mid [x]) < \alpha$, decide $BND(A)$

When $\alpha = \beta$, we have $\alpha = \gamma = \beta$. In this case, we use the decision rules:

(P2) If $P(A \mid [x]) > \alpha$, decide $POS(A)$,
(N2) If $P(A \mid [x]) < \alpha$ decide $NEG(A)$
(B2) If $P(A \mid [x]) = \alpha$, decide $BND(A)$

For the second set of decision rules, we use a tie-breaking criterion so that the boundary region may be non-empty.

DTRS model can offer new insights into the probabilistic approaches to rough sets. And it can generalize previous approaches to rough set models, including Pawlak rough set model and VPRS model. It is also noticed that DTRS model is obviously the underlying basis for a theory three-way decisions.

Now, we relate DTRS model to three-way decisions. Given a pair of thresholds (α, β) with $0 \leq \beta < \alpha \leq 1$, the sets of designated values for acceptance and rejection are $L^+ = \{a \in [0, 1] \mid \alpha \leq a\}$ and $L = \{b \in [0, 1] \mid b \leq \beta\}$. A DTRS model makes the following three-way decisions: for $A \subseteq U$,

$$
\begin{aligned}
POS_{(\alpha,\beta)}(A) &= \{x \in U \mid v_A(x) \succeq \alpha\} \\
&= \{x \in U \mid Pr(A \mid [x]) \geq \alpha\}, \\
NEG_{(\alpha,\beta)}(A) &= \{x \in U \mid v_A(x) \preceq \beta\} \\
&= \{x \in U \mid Pr(A \mid [x]) \leq \beta\}, \\
BND_{(\alpha,\beta)}(A) &= \{x \in U \mid \beta \preceq v_A(x) \preceq \alpha\} \\
&= \{x \in U \mid \beta < Pr(A \mid [x]) < \alpha\}.
\end{aligned}
$$

Three-way decision-making in DTRS can be easily related to incorrect acceptance error and incorrect rejection error. Specifically, incorrect acceptance error is given by $Pr(A^c \mid [x]) = 1 - Pr(Aj[x]) < 1 - \alpha$, which is bounded by $1 - \alpha$. Likewise, incorrect rejection error is given by $Pr(A \mid [x]) \leq \beta$, which is bounded by β. Therefore, the pair of thresholds can be interpreted as defining tolerance levels of errors.

Three-way decision-making in DTRS can be easily related to incorrect acceptance error and incorrect rejection error. DTRS model has an advantage that is based on Bayesian decision theory. In addition, the pair of thresholds can be systematically computed by minimizing overall ternary classification cost.

We can also deal with *fuzzy sets* of Zadeh [19] in a theory of three-way decisions. Recall that a fuzzy set $\mathcal{A}$ is defined by a mapping from U to the unit interval, i.e., $\mu_{\mathcal{A}} : U \to [0, 1]$. The value $\mu_{\mathcal{A}}(x)$ is called the degree of membership of the object $x \in U$.

Three-valued approximations of a fuzzy set can be formulated as follows. Given a pair of thresholds (α, β) with $0 \leq \beta < \alpha \leq 1$, one can define the designated sets of values for acceptance and rejection as $L^+ = \{\alpha \in [0, 1] \mid \alpha \leq a\}$ and $L^- = \{n \in [0, 1] \mid b \leq \beta\}$.

By Definition 11.3, if a fuzzy membership function $\mu_{\mathcal{A}}$ is used as an acceptance-rejection evaluation, i.e., $v_{\mu_{\mathcal{A}}} = \mu_{\mathcal{A}}$, the following three-way decisions are obtained.

$$POS_{(\alpha,\beta)}(A) = \{x \in U \mid v_{\mu_{\mathcal{A}}}(x) \succeq \alpha\}$$
$$= \{x \in U \mid \mu_{\mathcal{A}}(x) \geq \alpha\}$$
$$NEG_{(\alpha,\beta)}(A) = \{x \in U \mid v_{\mu_{\mathcal{A}}}(x) \preceq \beta\}$$
$$= \{x \in U \mid \mu_{\mathcal{A}}(x) \leq \beta\}$$
$$BND_{(\alpha,\beta)}(A) = \{x \in U \mid \beta \prec v_{\mu_{\mathcal{A}}} \prec \alpha\}$$
$$= \{x \in U \mid \beta < \mu_{\mathcal{A}}(x) < \alpha\}$$

Zadeh [19] gave the following interpretation of the above three-valued approximations of a fuzzy set:

(1) x belongs to $\mathcal{A}$ if $\mu_{\mathcal{A}}(x) \geq \alpha$,
(2) x does not belong to $\mathcal{A}$ if $\mu_{\mathcal{A}}(x) \leq \beta$,
(3) x has an indeterminate status relative to $\mathcal{A}$ if $\beta < \mu_{\mathcal{A}}(x) < \alpha$.

This interpretation is based on the notions of acceptance and rejection and is consistent with three-way decisions.

As discussed above, a theory of three-way decisions can generalize many set theories and many-valued logics. One accepts or rejects an object when its values from evaluations fall into some designated areas; otherwise, one makes a decision of noncommitment. And three classes of evaluations are developed.

We address that a theory of three-way decisions is very flexible to generalize existing theories. And we believe that a theory of three-way decisions is very promising as a basis for soft computing alternative to rough set theory and other rival theories.

References

1. Akama, S., Murai, T., Kudo, Y.: Reasoning with Rough Sets. Springer, Heidelberg (2018)
2. Chouchoulas, A., Shen, Q.: Rough set-aided keyword reduction for text categorization. Appl. Artif. Intell. **15**, 845–873 (2001)
3. Dubois, D., Prade, H.: Rough fuzzy sets and fuzzy rough sets. Int. J. Gen. Syst. **17**, 191–209 (1989)
4. Duda, R., Hart, P.: Pattern Classification and Scene Analysis. Wiley, New York (1973)
5. Grecco, S., Matrazzo, B., Slowinski, R.: Rough sets theory for multi-criteria decision analysis. Eur. J. Oper. Res. **129**, 1–47 (2001)
6. Hu, K., Diao, L., Lu, Y., Shi, C.: A heuristic optimal reduct algorithm. In: Proceedings of IDEAL2000, pp. 139–144. Springer, Heidelberg (2000)
7. Hu, K., Wang, G., Feng, L.: Fast knowledge reduction algorithms based on quick sort. Rough Sets and Knowledge Technology, pp. 72–79. Springer, Heidelberg (2008)
8. Kryzskiewica, M.: Rough set approach to incomplete information systems. Inf. Sci. **112**, 39–98 (1998)

 9. Kudo, Y., Murai, T.: Heuristic algorithm for attribute reduction based on classification ability for condition attributes. JACII **15**, 102–109 (2011)
10. Kudo, Y., Murai, T.: A parallel computation method for heuristic attribute reduction using reduced decision tables. JACII **17**, 371–376 (2013)
11. Lin, T.: Granulation and nearest neighborhood rough set approach. Granular Computing: An Engineering Paradigm, pp. 125–142. Springer, Heidelberg (2001)
12. Pagliani, P.: Rough sets and Nelson algebras. Fundam. Math. **27**, 205–219 (1996)
13. Pawlak, Z.: Rough sets. Int. J. Comput. Inf. Sci. **11**, 341–356 (1982)
14. Pawlak, Z.: Rough Sets: Theoretical Aspects of Reasoning about Data. Kluwer, Dordrecht (1991)
15. Skowron, A., Rauszer, C.M.: The discernibility matrix and functions in information systems. In: Slowinski, R. (ed.) Intelligent Decision Support: Handbook of Application and Advances of the Rough Set Theory, pp. 331–362. Kluwer, Dordrecht (1992)
16. Yao, Y.: An outline of a theory of three-way decisions. In: Proceedings of RSCTS 2012, pp. 1–17. Springer, Heidelberg (2012)
17. Yao, Y., Lin, T.: Generalization of rough sets using modal logics. Intell. Autom. Soft Comput. **2**, 103–120 (1996)
18. Yao, Y., Wong, S., Lingras, P.: A decision-theoretic rough set model. In: Ras, Z., Zemankova, M., Emrich, M. (eds.) Methodology for Intelligent Systems, vol. 5, pp. 17–24. North-Holland, New York (1990)
19. Zadeh, L.: Fuzzy sets. Inf. Control **9**, 338–353 (1965)
20. Ziarko, W.: Variable precision rough set model. J. Comput. Syst. Sci. **46**, 39–59 (1993)
21. Zhu, W.: Topological approach to covering rough sets. Inf. Sci. **177**, 1499–1508 (2007)

Index

Printed in the United States
By Bookmasters